Planes without Pilots

Advances in Unmanned Flight

Planes without Pilots
Advances in Unmanned Flight

Bill Siuru

TAB AERO
Blue Ridge Summit, PA

FIRST EDITION
FIRST PRINTING

Library of Congress Cataloging-in-Publication Data

Siuru, William D.
Planes without pilots : advances in unmanned flight / by Bill Siuru.
p. cm.
Includes index.
ISBN 0-8306-3432-0 (pbk.)
1. Drone aircraft. I. Title.
TL671.S5653 1991 90-20627
629.132′6—dc20 CIP

TAB Books offers software for sale. For information and a catalog, please contact TAB Software Department, Blue Ridge Summit, PA 17294-0850.

Questions regarding the content of this book should be addressed to:

Reader Inquiry Branch
TAB Books
Blue Ridge Summit, PA 17294-0850

Acquisitions Editor: Jeff Worsinger
Book Editor: Steven H. Mesner
Production: Katherine G. Brown
Series Design: Jaclyn J. Boone
Cover Design: Lori E. Schlosser
Front Cover Photograph: Boeing Brave-200 courtesy of Boeing Military Airplane Company.

Contents

Acknowledgments

This book could not have been written without the help of many organizations that provided information, illustrations, and photographs.

Aerospatiale SNI
Beech Aircraft Corporation
Boeing Aerospace Company
Boeing Canada, Ltd.
Boeing Helicopters
Boeing Military Aircraft Company
Cadillac Gage Textron
Canadair, Inc.
Defense Advanced Research Projects Agency
Developmental Sciences
E-Systems Inc.
Farner Air Service
Federal Aviation Administration Technical Center
Flight Refueling Ltd.
Ford Aerospace and Communications Corporation
General Atomics
Grumman Corporation
Israeli Aircraft Industries
Lockheed Aeronautical Systems, Georgia Division

Lockheed Austin Division
Lockheed Missile and Space Company
Martin Marietta Aero and Naval Systems
Martin Marietta Denver Division
Martin Marietta Orlando Aerospace
McDonnell Douglas
ML Aviation Ltd.
Mollier International
NASA/Dryden Flight Research Center
National Aeronautics and Space Administration
Northrop Corporation
Sandia National Laboratories
Standard Manufacturing Co., Inc.
Targetair, Ltd.
Target Technology Ltd.
Teledyne Ryan Aeronautical
United States Air Force
United States Army
United States Navy
USAF Aeronautical System Division
USS *Iowa*
Westland Helicopters Ltd.
Wint Clean AB
Julian Wolkovitch

Special thanks are given to Brian Siuru, who proofread the manuscript for this book.

Introduction

Remotely piloted vehicles, drones, unmanned air vehicles—or whatever their current title might be—are among the most exciting and important craft in today's skies. In the following pages we will look at UAVs in detail, both past and future. This book describes not only the UAVs, but the technologies that make them possible as well as the jobs and roles they are best suited to perform.

Hopefully, the book will provide a better understanding of UAVs and help fill in the gaps in stories that will be making the news in the future. As you will see, unmanned aircraft have had their ups and downs depending on current military, political, and budgetary environments.

Also presented are the little-known facts about remotely-piloted vehicles and drones that have already earned their battle stripes in World War II, in the Middle East, and especially during the conflict in Southeast Asia. This book also touches on the many attempts that never got much beyond the drawing board or demonstration flight stages, for the UAV's history is filled with many more aborted programs than craft that actually saw operational service.

This book is for anyone who has an interest in unmanned aircraft whether an aviation buff, radio control model fan, aerospace historian, or just plain curious.

1

UAVs, RPVs, and Drones Defined

FUTURE wars could be fought with minimum "face-to-face" contact between opposing forces. Changes in demographics (such as a sizable reduction in the pool of youths of draft age) and political issues (such as those experienced during the Vietnam era) are causing military leaders to look for ways to extend the capabilities of each soldier, sailor, airman, and marine as well as remove much of the risk to human life.

America's economic conditions will require less expensive weapons in the future, and unmanned aircraft can be much less expensive than manned ones. Technological advances in such areas as optics, sensors, microelectronics, computers, composite materials, robotics, and artificial intelligence are now making unmanned air vehicles possible. But besides their applications as weapons, unpiloted aircraft can be used for commercial and civilian jobs and for safely and inexpensively testing new aircraft concepts.

A Few Definitions

Aircraft without pilots are called by a variety of names and acronyms. *Unmanned Aerial* (or *Air*) *Vehicles* (UAV) or *UnManned Aircraft* (UMA) are perhaps the most generic terms. The term *Unmanned Vehicle System* (UVS) is another generic title that is less frequently used. *Remotely Piloted Vehicles* (RPV) implies that the aircraft is flown by a human pilot on the ground or in another aircraft. An RPV could be considered a sophisticated radio-controlled aircraft that uses the same basic techniques that are familiar to the R/C hobbyist.

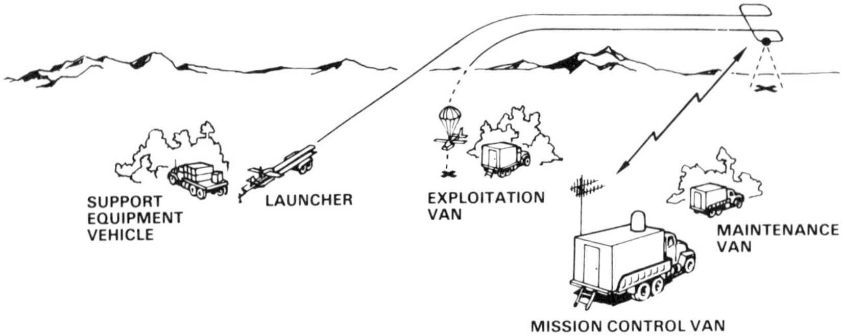

Fig. 1-1. Typical nonlethal UAV mission profile. In this case, the UAV is launched for reconnaissance behind enemy lines and recovered by parachute.
E-Systems

Many military UAVs are designed to fly on a preprogrammed course with remote commands given only periodically to update the flight or make corrections. Autonomously flown aircraft are frequently referred to as being "fire-and-forget" weapon systems. UAVs includes *drones*, a title commonly used when referring to unpiloted aircraft used for target practice, but which also includes any unpiloted aircraft that flies with little or no human assistance. Weapons such as bombs, projectiles, and rockets that are guided while in flight have received the title "smart" weapons and fall within the broad definition of UAVs. So do ballistic, surface-to-air, air-to-air, and air-to-surface missiles. Similarly, space probes such as the *Magellan* and *Galileo* (which were launched in 1989 for extended trips to Venus and Jupiter, respectively) fit the generic term, but are not part of this story.

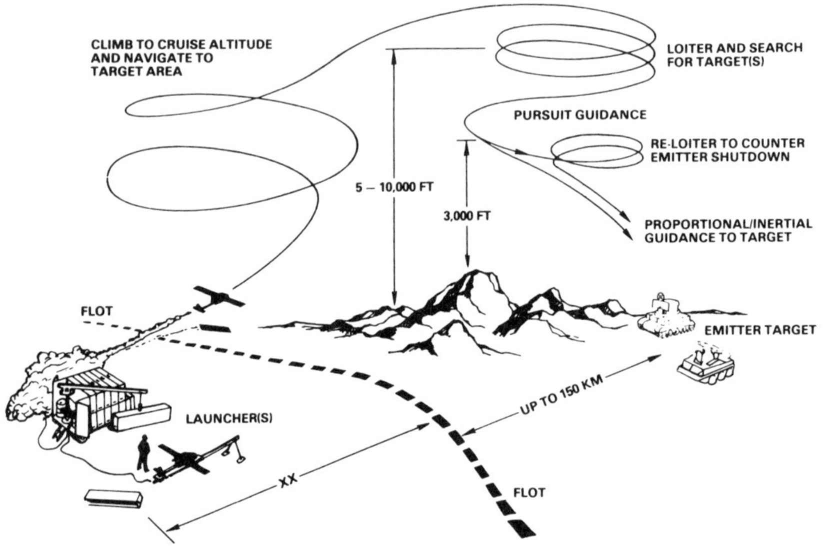

Fig. 1-2. Typical lethal UAV mission profile. In this case, the UAV is used as a tank destroyer. E-Systems

Some of the techniques and technology used in smart weapons, missiles, and space probes are often found in the UAVs we will be discussing. Similarly, much of the technology used in remotely controlled ground vehicles [such as *Robotic Combat Vehicles* (RCV) of interest to the ground forces] is applicable to military UAVs. This is especially true because both must be simple, rugged, and able to survive in battle.

Unmanned vehicles have been used to flight-test new aeronautical concepts; these UAVs are appropriately called *Remotely Piloted Research Vehicles* (RPRV). Finally, because many of the UAV concepts are rotary-wing craft or helicopters, the term *Remotely Piloted Helicopter* (RPH) has also been coined.

Why UAVs?

For decades, interest in unpiloted aircraft has waned and flourished in turn. Historically, UAVs have been given low priority during peacetime, partly perhaps because the technology was less exciting to the engineer and scientist and partly because many decision-makers in the military arena have been pilots who felt more comfortable flying from the cockpit. After all, money spent on UAVs meant less funding for new *manned* aircraft, which are a lot more fun to fly. Unmanned aircraft are also perceived to be less "career-enhancing" for the military officer. The fact that some previous UAV programs have grown immensely in terms of cost and complexity and still failed to meet their original objectives has not helped the UAV's case either.

However, in wartime, remotely piloted vehicles have repeatedly shown that they can perform dangerous missions without putting a human pilot or a costly aircraft in jeopardy. Today the future of military UAVs is much more promising for several reasons.

Military experts tell us that if we should ever have to fight the Soviets and Warsaw Pact nations in a conventional war, NATO forces would be outnumbered 7 to 1 in air-defense missiles, at least 2 to 1 in tanks, artillery, and fighter aircraft, and 3 to 2 in troops. To date, our chief counter to these overwhelming odds has been tactical nuclear weapons based in Europe.

Now there is a willingness on the Soviets' part to negotiate the removal of intermediate-range nuclear missiles from Europe. If negotiations to eliminate these weapons succeed, the NATO nations will have to find a new, conventional means to substitute for the tactical nuclear umbrella that counters the Soviet and Warsaw Pact numerical superiority in troops, planes, tanks, missiles, and guns. The West still retains one important advantage, its lead in technology. We are still well ahead in such key areas as computers, microelectronics, robotics, artificial intelligence, and optics. Thus the military-industrial complex is investing huge sums of money in smart weapons, advanced sensors, unmanned air vehicles, and battlefield robots to replace sheer firepower and manpower with electronic and computer power.

Military planners see an alarming trend when they look at demographic predictions. By 1995, the pool of 17-to-21-year-olds will drop by nine percent—that is, about 1.8 million fewer men and women for the recruiters to work with. In West Germany, our NATO partner with the largest army in Europe and the lowest

Ford Aerospace and Communications Corp.

Fig. 1-3. The Prairie IIB mini-RPV, designed in the 1970s, is an example of a non-lethal UAV that works with a lethal weapon. The Prairie carried a television camera and a laser target designator. The latter illuminated the target (such as a tank) with a laser beam. A "smart" weapon, such as the Army's Cannon-Launched Guided Projectile (CLGP) fired from a 155mm howitzer, would home in on the laser reflection from the target to destroy it. Like so many UAVs, the Prairie never reached production.

birth rate in the world, things look even more dismal, with the draftee pool being cut in half by the mid-1990s.

The answer must be to do more with each combat troop. Lt. Gen. William Thurman, when he was the Commander of the USAF's Aeronautical Systems Division, said that "We need two of our guys to equal ten." Thus, future weapons—right down to those of the individual soldier—will have to be *force multipliers*. Each will have to allow a single troop to do the job of two, three, or even five individuals. Furthermore, from our past experience (especially in Southeast Asia), it is doubtful that the American people will ever again allow wholesale commitment of American troops to another armed conflict, whether it be large or small.

Unmanned air vehicles have another important advantage. They can be significantly cheaper to develop, build, and use compared to manned-rated aircraft that require sophisticated life-support systems. UAVs can be especially inexpensive if they are expendable. Not only would such UAVs be cheap enough that they could be fielded in plentiful quantities, but once launched they could be forgotten. Military combat leaders especially like a "fire-and-forget" capability because it frees ground troops of the duties of flying and recovering RPVs during the heat of battle. Instead, they can be launching more unmanned aircraft.

UAVs are ideal for hazardous missions such as reconnaissance behind enemy lines or destroying heavily defended targets. As the enemy gets more sophisticated and capable sensors such as radar and infrared detectors, more and more missions fall into the "extremely dangerous" category.

A major factor for UAV successes is that the technology is now ready. As an aside, from the engineer's and scientist's standpoints, the applications have now become interesting and challenging as well. Technologies such as *Very High Speed Integrated Circuits* (VHSIC), artificial intelligence, optical sensors, and robotics are mature enough to be adapted to low-cost, field-ready applications not only in UAVs, but battlefield robots and smart weapons.

Finally, there is currently an impetus being applied by Congress for the development of UAVs. For instance, Congress recently required the U.S. military to prepare a joint service master plan for the development of UAVs. This is important because it integrates all the services' non-lethal UAV programs into a single joint service program office with the Navy being designated as the executive service. The joint service setup is designed to prevent duplicate and overlapping UAV programs. This was at least a perceived and sometimes a real problem in the past, when each service was independently developing UAVs for their own needs. Perhaps most important, Congress has allocated funding for developing UAVs in the 1990 to 1994 time period, including funds to determine the feasibility of establishing a common site for UAV testing, evaluation, and training.

Robotic systems of all types are not without their problems. Of course there are political and moral questions, such as: "Will unmanned systems encourage war because human life is less likely to be at stake," or "How will robots take prisoners?" These are important issues for the tacticians and technologists.

One of the key challenges is to keep the UAVs as simple as possible. First, if they become too sophisticated, they also become too expensive to purchase and use, especially for vital training. Such unbounded growth in complexity usually results when military planners keep demanding more capability. This growth has already caused the demise of several originally attractive remotely piloted vehicles. Finally, any unmanned vehicle has to be operated by the *average* soldier under *combat* conditions, not an electrical engineer in the comfort of a laboratory. Ideally, systems with sophisticated electronics inside should still be as "user friendly" as an M16 rifle.

2

A Bit of History

UNMANNED aircraft concepts have been around almost as long as the airplane itself. The earliest concepts were for either unmanned targets or weapons of destruction. In the latter case, cheap, expendable aircraft were planned for carrying bombs without risking a human pilot on dangerous missions. Most of these never went beyond the idea stage (or, if they were a bit more fortunate, the prototype stage) for the same reason as many of the more recent UAV concepts: lack of funding.

Before World War II

Lawrence Sperry invented the gyrostabilizer and installed it in a Curtiss biplane in 1913. His invention led to the autopilot for manned aircraft and was also the foundation for unmanned ones. By 1917, the invention was installed in a Sperry-Curtiss "aerial torpedo" that was successfully flown several times over Long Island, New York.

During the same period, Charles "Boss" Kettering, the brilliant inventor who brought us the electric automobile starter and tetraethyl lead, which made the high-compression automotive engine feasible, was working on his Kettering Aerial Torpedo. This UAV, better known as the Kettering "Bug," was built by the Dayton-Wright Airplane Company in 1918 for the U.S. Army Signal Corps. Although it proved successful in flight tests, World War I ended before it saw combat. Its development was halted by the early 1920s because of the scarcity of funding, and only about 100 Bugs were built.

The Bug was just over 12 feet long and had a wingspan of 15 feet. Construction materials included wood, muslin, butcher paper, and pasteboard to keep the craft *very* inexpensive. The Bug took off from a dolly that ran along a track to achieve takeoff speed. It was guided toward its target by preset vacuum-pneumatic and electric flight controls. Once the desired distance had been covered (typically 30 to 40 miles), the Bug's wings were ejected and the noseheavy fuselage then dived towards the Earth with a payload of about 200 pounds of explosives. A gyroscope and altimeter kept the craft flying at the proper altitude and in the right direction. Distance was measured by a propeller and a "subtracting" counter. When the desired number of engine revolutions had been subtracted from the preset number of total turns, an electric circuit was energized to jettison the wings and shut off the 40-horsepower, two-cycle, four-cylinder engine.

Meanwhile, in England, Professor A.M. Low was designing another remotely controlled bomb. Codenamed the "Aerial Target," the weapon was aimed at destroying both ground targets and German Zepplins. While prototypes were built by de Havilland and the Royal Aircraft Factory, they too did not see action in the war.

In the 1920s, the British developed one of the earliest (if not *the* earliest) truly remotely controlled aircraft. Unlike predecessors that used an onboard mechanical flight control system to fly a preprogrammed flight plan, this RPV was flown by radio control. The RPV took off from the deck of the HMS *Stronghold* in September 1924 on a 12-minute flight that was terminated because of an engine failure. In the next few years, the British continued a successful pilotless aircraft program. For instance, LARYNX (for *Long-Range gun with Lynx Engine*) successfully flew a 300-mile course in July 1927, again after launch from the *Stronghold*. After this, the British program was directed toward target drones.

In the United States there were several aborted unpiloted aircraft ventures between the wars. For instance, around 1928, a Curtis Robin monoplane was fitted with radio controls and bombs; it flew off-and-on as funds were available until 1932, when there were no funds at all.

In the late 1930s, as war clouds loomed on the horizon, the military again became interested in unmanned aircraft. The Army started several "special weapons" projects. These included a series of PQ radio-controlled target planes, a GB-1 glide-bomb, the AZON controllable high-angle bomb, a surface-to-surface "buzz" bomb called the General Motors "Bug," and another glide-bomb dubbed the "Bat." Charles Kettering was again involved in many of these projects.

World War II

One of the most successful of these special weapons was the GB-1, which actually saw combat in World War II, albeit on a rather limited scale. The GB-1 was essentially a 2000-pound bomb fitted with 12-foot long wings and twin rudders, all made of plywood. The GB-1 was carried aloft by a B-17 and dropped. The bomb was then "flown" to its target via radio control by a "pilot" aboard the B-17. The GB-1 allowed bombing to be done while the B-17 was well out of the range of concentrated antiaircraft fire, but the bomber did have to stay within a range where the GB-1 could be visually guided to its target. The GB-1 was used on at least one combat mission in 1943, when 54 B-17s, each loaded with two GB-1s, bombed Cologne, Germany.

During World War II, the U.S. also attempted to use radio-controlled worn-out B-17s, B-24s, and even B-29s filled with explosives to destroy enemy targets. Takeoff was accomplished by a human crew, who bailed out once the aircraft was airborne and on course; it was then radio controlled to the target. Navy Lieutenant Joseph P. Kennedy lost his life in one of the Project Aphrodite B-24 bombers when things went wrong before he could bail out. The cost of preparing the war-weary bombers turned out to be quite expensive and there were often problems with the radio control system.

One of the most successful unmanned weapons of World War II was the Navy's Bat, which appeared in the Pacific Theater during the last year of the war. The Bat was a glide-bomb that homed in on its target using radar. This was a low-wing monoplane with a plywood airframe and a 1000-pound bomb in the fuselage. The Bat was launched from a PB4Y-2 Privateer patrol bomber well outside the range of antiaircraft fire (the Bat's range was about 20 miles) and flown using its gyropilot, radar, and aerodynamic controls. Bats were credited with sinking many tons of Japanese shipping, including a Japanese destroyer. They were also used against land targets such as bridges.

The Germans also developed and used unmanned aircraft. If the V-1 "buzz bomb" and the V-2 rocket were counted as UAVs—and they should be—you would have to say that the Germans were significantly more successful than other WWII combatants. The German Ruhrstahl SD.1400X, often referred to as the X-1 or Fritz X, was a radio-controlled bomb with control surfaces that could be moved to guide the weapon to its target. The bomb was controlled from a "mothership" bomber. This weapon is credited with sinking the Italian battleship *Roma*. The Germans also used the Henschel Hs 293A, which was a rocket-powered flying bomb. Besides using UAVs for mass destruction, the Germans experimented with V-series missiles fitted with cameras for reconnaissance.

However, the UAV's real contribution to the war effort was as a target for training airborne and ground-based gun crews. During the 1930s, the British learned much about both gunnery and RPVs in its highly successful target drone program. In the United States, many aircraft such as Tiger Moth biplanes were converted to fly pilotless for target practice. A Hollywood actor by the name of Reginald Denny produced a target drone with a nine-foot wingspan that proved to be a less costly alternative to the rather expensive converted manned aircraft. During World War II, the U.S. Army and Navy used nearly 14,000 drones for training.

After the War

After World War II, interest in RPVs was continued on a limited scale but they were seldom heard about. Their stories were not particularly glamorous when compared to the manned aircraft advances that were making the headlines.

One of the more successful UAVs of the 1950s and 1960s were the RPHs

(*Remotely Piloted Helicopters*) developed by the Gyrodyne Company of America, Inc. These craft initially used two coaxial rotors for lift and one or more propellers for thrust. The first aircraft built using this design was the tiny Model 2B, which was flown in 1949. This was followed by the XRON-1 X Rotorcycle in 1955. Both of these were manned helicopters. In 1958, Gyrodyne was awarded a contract from the Navy to modify the concept so it could be flown remotely from the deck of a destroyer carrying a torpedo as part of the DASH (*Drone Anti-Submarine Helicopter*) system.

Over the next decade, nearly 800 QH-50s, the official designation for the Gyrodyne-built UAV, were built in at least four versions, along with several prototypes for testing new ideas. While the earliest variants, the QH-50A and B, were powered by 72-horsepower Porsche engines, the larger QHC-50C/D used a 300- to 330-horsepower Boeing turboshaft engine. The QH-50 was flown by radio control using both human vision and radar to keep track of the craft's flight. While primarily designed for operation off a ship, the QH-50 could also be controlled from another manned aircraft or a mobile ground vehicle.

US Navy

Fig. 2-1. *The QH-50 was the flying portion of the Navy's DASH anti-submarine warfare system and saw limited duty in Southeast Asia.*

These early UAVs were truly multipurpose craft and were used to demonstrate a wide variety of capabilities besides carrying torpedos and surveillance during ASW (*Anti-Submarine Warfare*) missions. The QH-50s demonstrated that unmanned helicopters could drop sonobuoys and flares, perform rescues, transport cargo, illuminate targets at night, and lay down smokescreens.

Even more RPH capability was demonstrated during *Project Snoopy*, *Nite Panther*, and *Nite Gazelle* test programs conducted during the Vietnam era. In *Project Snoopy*, QH-50C/Ds were equipped with TV cameras that were used for surveillance and target spotting. The TV pictures were transmitted to a destroyer and used to aim the shipboard guns at targets on the coast of Vietnam. In *Nite Panther*, the QH-50 was fitted with a variety of sophisticated equipment including day/night TV cameras, motion and still cameras, moving target indicator, laser rangefinder/target designator, and a covert illuminator. *Nite Gazelle* took the *Nite Panther* project a step further by adding weapons including a minigun, high-velocity gun, grenade launcher, bomblet dispenser, bombs, and a laser-guided rocket.

Even after the Navy mothballed the QH-50s, the most widely used rotary-wing UAV to date, its life was not over. The QH-50 was used in both the Aerodyne Systems CH-84 Pegasus and Dornier Argus-Priamos UAV programs that were underway as into the late 1980s.

After World War II, the primary interest in UAVs continued in the form of a multitude of guided missiles of all types that were developed and actually put into operation through the years. There are too many to give a complete history here, but the significant ones will be briefly summarized.

The Boeing-built Bomarc was a winged interceptor missile that used a rocket booster and two ramjet engines to intercept enemy aircraft at distances of up to 400 miles and altitudes of up to 100,000 feet. The Bomarc was guided from the ground until it neared its target, at which point a target seeker took over to home the missile in for the kill with its nuclear warhead. The Bomarc, whose production ended in 1962, was used by both the United States and Canadian Air Forces.

USAF

Fig. 2-2. Guided missiles such as the USAF Mace are a form of UAV and much of the same technology is applicable to the more traditional UAVs.

The MGM-13 Mace and its predecessor, the Matador, were built by the Martin Company. These rocket-boosted and turbojet-powered missiles were launched from a roadable launcher and used either a map-matching or inertial guidance system. They had a range of 650 to 1200 miles carrying a nuclear weapon against a ground target. The AGM-20 Quail, built by McDonnell Aircraft, was launched from a B-52 to confuse enemy radar. Other missiles such as the Falcon, Sparrow, and Sidewinder, used onboard radar or infrared detectors to home in on the target. Still other missiles, such as the Bullpup, TOW, Maverick, and Stinger are "flown" to the target by a remotely located human "pilot." Much of the technology from these unmanned weapon systems is applicable to the unmanned vehicles such as remotely piloted and robotic aircraft.

The Firebee

No story about pilotless flight would be complete without including the huge family of UAVs built by Teledyne Ryan Aeronautical, formerly Ryan Aeronautical Company. For decades, Ryan has been producing its Firebee jet-engined UAVs. The first in the series, the Model 124 Firebee I, was designed as a target drone and first flew in 1951. The Firebee has been used as a target drone by all three military services and several foreign nations. Well over 6500 have been built. The Firebee was also built under license in Japan by Fuji Heavy Industries. The Air Force's version is designated the BQM-34A, the Army's is the MQM-34D and the Navy uses the BQM-34A and the BQM-34S.

The Ryan Firebee can be launched from beneath a specially modified aircraft, from a ground launching platform, or from a ship. Surface launching is assisted by a solid-propellant rocket and flight is sustained by a turbojet engine. A parachute is used to recover the craft. The Firebee target drone can be controlled from either the ground or from a manned aircraft. Besides being used for training surface-to-air and air-to-air gunnery and missile launch crews, the Firebee has been used in developing and testing new weapon systems. There probably is not a U.S. surface-to-air or air-to-air missile that hasn't been tested against the Firebee.

In the late 1960s, the Firebee II, still based on the Firebee I, was developed as the BQM-34F for the USAF and BQM-34 and BQM-34T for the U.S. Navy. This target drone could operate at altitudes of up to 60,000 feet and could travel at speeds of up to Mach 1.5 for 14 seconds. By comparison, the Firebee I target drones were subsonic with a maximum speed of 600 knots (690 mph).

The UAV concept's finest hour to date has to be the Firebee's successful use in Southeast Asia. These were Ryan Model 147 Firebee drones, often called "Lightning Bugs." The Model 147 RPV was derived from the Model 124 Firebee I target drone. There were over 30 versions of the Model 147, all powered by single turbojet engines. The original Model 147A was developed in the early 1960s to fill a void in unmanned reconnaissance capability exposed by the downing of Gary Powers' U-2 over the Soviet Union, the lost of an RB-47 due to Soviet action in the Bering Sea, and the Cuban Missile Crisis.

Thus the first of the high-altitude Lightning Bugs, Model 147B, were ready

Teledyne Ryan Aeronautical

Fig. 2-3. *One of the world's most popular target drones, the Ryan BQM-34A Firebee I. While normally air-launched, this version was ground-launched using a solid propellant JATO bottle for the initial boost.*

to perform photo reconnaissance missions with U.S. involvement in Vietnam triggered by the Tonkin Gulf incident of August 4, 1964. Within the month, the Lightning Bugs were in operation overflying the Chinese mainland to obtain intelligence data on Communist China's progress on A-bomb development and missile deployment, as well as industrial complexes and military installations.

The Model 147Bs, like most of the Ryan-built drones, were carried on pylons beneath a specially equipped C-130 Hercules transport (DC-130). The flights originated out of Kadena AB on Okinawa, with the drones released around Hong Kong; recovery via parachutes occurred on Taiwan. To thinly disguise U.S. involvement, the drones were painted with Nationalist Chinese insignia. (Incidentally, these were *drones* rather than *remotely piloted vehicles*, since each flight was entirely preprogrammed, with no control from either the DC-130 mothership or the ground.)

The Model 147Bs made the news when the Chinese Communists reported in 1964 that they had shot down a pilotless American reconnaissance plane over south central China. This was followed by more reported downings, plus a display of the wreckage of at least three at a Chinese museum in Beijing.

Teledyne Ryan Aeronautical

Fig. 2-4. The Teledyne Ryan Model 147SD was used for low-altitude, real-time surveillance missions.

By late 1964, the Lightning Bugs were in operation in Vietnam. The 100th Strategic Reconnaissance Wing operated out of Bein Hoa with recovery in the Da Nang area. Between August 1964 and June 1975, the 100th SRW flew 3435 operational sorties. Because so many 147s were damaged due to hard landings, the mid-air retrieval system or MARS was developed. With MARS, the craft were snatched in midair by their parachutes with helicopters.

While there were about 30 Model 147 variants, by far the most numerous version was the Model 147SC, which normally flew at low altitudes, taking photographs with detail not usually available from manned reconnaissance aircraft. The Model 147SC had remarkable recovery rates. Of the 1651 missions flown by several hundred SCs between January 1969 and June 1973, mainly over heavily defended North Vietnam, the drones were recovered over 87 percent of the time.

Teledyne Ryan Aeronautical

Fig. 2-5. The DC-130 "motherships" could carry up to four Firebee UAVs.

The more sophisticated Model 147SD low-altitude reconnaissance craft, which provided real-time intelligence data, had the best survival rate—97.3 percent. The average drone flew ten missions before it failed or was shot down.

Other versions were built to perform electronic intelligence, electronic countermeasures, communication intelligence, and covert psychological warfare missions. For instance, the Model 147NC was used to drop propaganda leaflets on North Vietnam during *Project Litterbug*.

During the *Rolling Thunder* bombing campaign over North Vietnam from 1966 through 1968, Lightning Bugs were used to collect targeting data and do post-attack damage assessment for the raiding B-52s. They were able to cover areas such as Haiphong and Hanoi denied to manned reconnaissance aircraft because of the high concentration of enemy air defenses. Peak business for the SC birds was during the period of 20 December 1972 and 19 January 1973, when over 100 sorties were flown mostly on BDA missions. During the 11 days of *Linebacker II*, the drones flew 27 missions in a one-week period.

Teledyne Ryan Aeronautical

Fig. 2-6. Tom Cat, *a model 147SC, flew a record 68 sorties before it was finally lost.*

One of the drones' important accomplishments was locating prisoner of war camps so they would not be accidentally bombed by U.S. warplanes. For instance, the existence of the Son Tay prison was confirmed by photos taken from a reconnaissance drone.

Another variant, the Model 147N, was used as a decoy to confuse enemy air defense radars while the Model 147Gs flew their high-altitude spy missions. These decoys did not have parachutes because they were not expected to survive. The N birds are even credited with five "unofficial" MiG kills of enemy fighters that ran out of fuel or were mistakenly shot down by North Vietnamese SAMs or other MiGs. Because some N models did survive and return to fly around Da Nang until they ran out of fuel, the NX model was fitted with a parachute recovery system and an inexpensive camera, although its primary mission was still as a decoy.

After the cessation of bombing raids in early 1973, the Firebees' task got even tougher, because now the MiGs and SAMs could concentrate on the drones. The Firebees and the SR-71 were the only aircraft allowed to overfly the North.

During the Vietnam era, there were other uses for the Ryan RPVs that were tested—sometimes under combat conditions—but never put into full-scale operation. For example, Ryan built 10 Model 147SKs for the Navy. These reconnaissance RPVs were modified so they could be launched from the deck of an aircraft carrier using a JATO (*Jet Assisted Takeoff Bottle*) for extra takeoff boost. The Model 147SK was tested under combat conditions over Vietnam and one was shot down over China. These were launched from the attack carrier USS *Ranger*.

The Navy was interested in the FLASH (*Firebee Low Altitude Ship-to-Homing*) missile to defend its ships at sea against potential attacks from enemy missile-carrying ships. Like the Model 147SK, the FLASH used a JATO bottle for takeoff. It was remotely guided from aboard the ship or a helicopter overhead using radio control and a TV camera located in the nose of the aircraft. There was also a warhead in the nose. While the system worked well in tests, the Navy elected to wait for the Harpoon anti-shipping missile, which was in development.

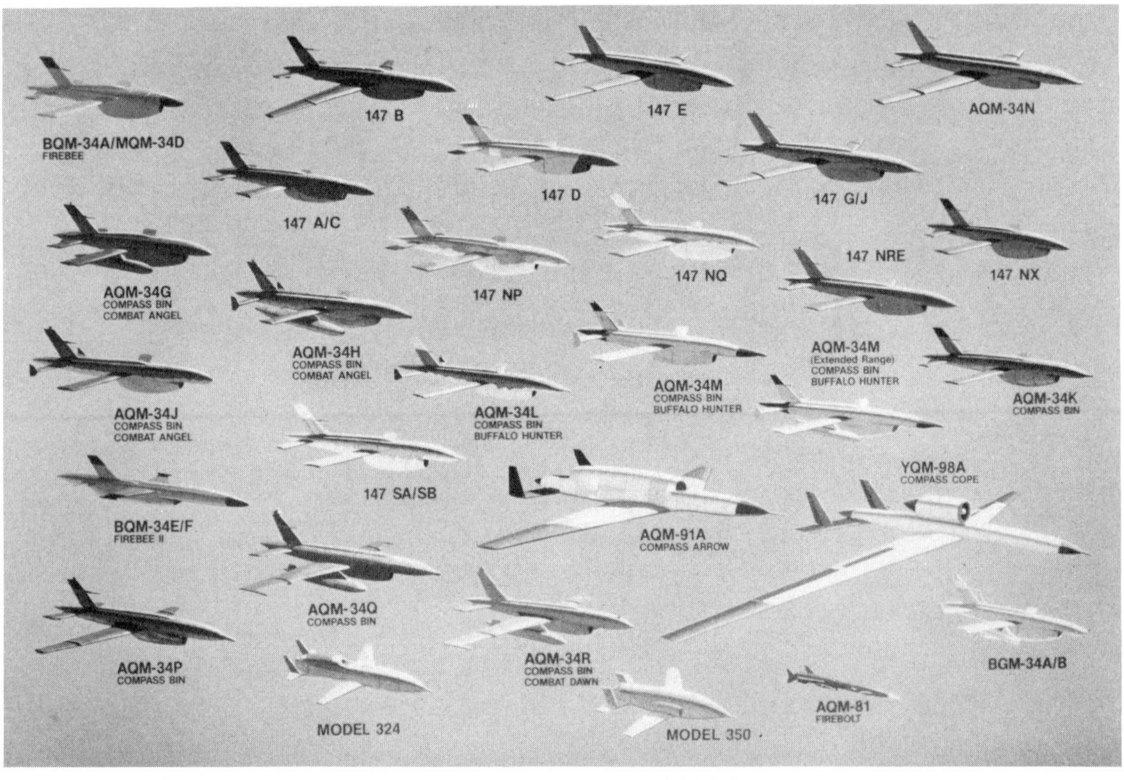

Teledyne Ryan Aeronautical

Fig. 2-7. *The family of UAVs based on the Teledyne Ryan Model 147 drone.*

Firebees were also developed during this era to carry weapons. For example, Ryan modified an RPV so it could carry weapons such as the Maverick optical-seeking missile and HOBO "smart bomb" to be used as a defense suppression weapon to destroy enemy radar installations. After launch from the DC-130, the missile-carrying RPV would be remotely guided to the target area via TV onboard the drone. The idea was to use the RPV as an anti-radar weapon to soften up the target so fighters could go in to finish the job with reduced risk. While the system worked well, the SAM sites in North Vietnam were so well camouflaged that the human eye could not find them, much less a TV camera or optical tracking weapon.

The Model 147NA was fitted with twin ALE-2 pods that were chaff dispensers or active ECM (electronic countermeasures) equipment to jam enemy radars. The Model 147NA was assigned to TAC's 11th Tactical Drone Squadron.

Just as so often in the past, at the end of hostilities, the USAF mothballed its fleet of Model 147 Firebees. Ryan (now Teledyne Ryan) remained in the RPV business and has produced other RPVs and developed numerous prototypes.

The Model 234 was developed to carry weapons such as Shrike and Maverick missiles. The Model 234A carried *Low-Light-Level TV* (LLLTV) or *Forward-Looking InfraRed* (FLIR) sensors and a laser system to locate and designate targets for other RPVs carrying weapons. The Model 239 evaluated equipment for performing reconnaissance, electronic warfare, and air-to-ground strike missions. The Model 154 (AQM-91A) with its "stealth" characteristics was designed for performing overflights of China after earlier drones had been shot down.

The Model 235 R-Tern (YQM-98A) was Teledyne Ryan's contender in the USAF's Compass Cope high-altitude, long-endurance RPV program, which was cancelled in 1977. Boeing, with its YQM-94A B-Gull RPV, was the other contender—and the winner—in this competition. This RPV was intended for battlefield, signal intelligence, communications relay, ocean surveillance, and atmospheric sampling missions. Both the Ryan and Boeing versions could operate at altitudes of over 55,000 feet for periods in excess of 20 hours.

Other high-altitude RPVs developed by Ryan included the Model 305 Firebolt high-altitude supersonic target (HAST) with top speed of Mach 4, and the Spirit for the high-altitude long-endurance mission (as high as 75,000 feet and 80 hours). Ryan has also developed mini-RPVs such as the Model 262 and the Model 328. The latter, designed for airfield damage assessment, weighed only 258 pounds and was powered by a 25-horsepower two-cylinder gasoline engine.

Other U.S. RPV Builders

Beech Aircraft is another American company with extensive experience in target drones, having built over 7000 of its MQM-39A, MQM-61A, AQM-37, and MQM-107 target drones for the U.S. Air Force, Army, and Navy since the mid-1950s. The Beech Model 997 has been selected by the U.S. Navy as a replacement for the venerable Ryan BQM-34 Firebee.

Another American company with much RPV experience is Northrop, whose Basic Training Targets (BTT), the MQM-33 and MQM-36, have been in production since the mid-1940s, albeit with many changes and improvements. Indeed, the BTT probably holds the record as the aircraft produced in the greatest numbers with over 65,000 built, and further production is planned into the 1990s. Northrop has also built some 3200 of its MQM-74A and MQM-74C Chukar I

Fig. 2-8. The Beech AQM-37 has been another prolific target drone, with over 4000 built since it first flew in 1961. This supersonic drone with a Mach 3.0 (Mach 4.0 in extended performance version) capability has been used by the U.S. Air Force, Army, and Navy. Normally, it is not recovered.

and II target drones since 1964. Besides the U.S. Navy, the Chukar has been sold to several other countries including Israel, Britain, Italy, West Germany, Greece, Iran, Japan, the Netherlands, Saudia Arabia, and Spain. The Chukar III (BQM-74C) is the most recent version of the Northrop-produced drone, which will be produced into the 1990s. There is also a reconnaissance version of the Chukar, the BQM-74C/Recce, which uses a daylight TV camera.

The Israeli Connection

The Israelis are the most recent users of UAVs in combat—and highly successful ones, too. During the 1973 Yom Kippur War, Israel used "harassment" drones to

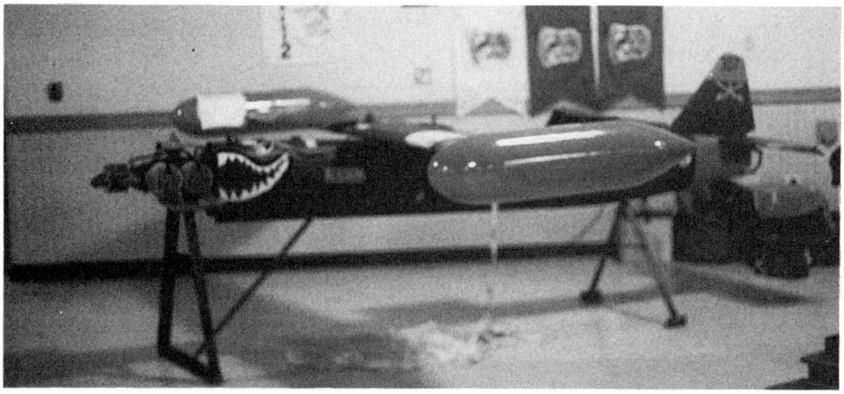

Fig. 2-9. The Northrop Basic Training Target (BTT) has been produced in greater quantity than any other aircraft, with at least 65,000 built since 1945.

to saturate enemy air defenses screening the flights of its fighter-bombers; these spoofed the Egyptians into wasting their surface-to-air missiles on the drones rather than the fighters. The drones used were low-cost Northrop-built Chukars. The Israelis also used the dozen or so Model 124L Firebee drones it had obtained from Teledyne Ryan for this purpose, as well as for day/night reconnaissance missions.

After their success in the 1973 October War, Israel established a flourishing RPV industry to build unmanned air vehicles for domestic use as well as for foreign sales. Initially, Israel Aircraft Industries (IAI) and Tadiran built the Scout and Mastiff mini-RPVs. In 1984, these two concerns consolidated their RPV

Israeli Aircraft Industries

Fig. 2-10. *The highly successful Israeli Scout mini-RPV.*

activities into Mazlat, Ltd. The Israelis have concentrated on mini-RPVs, such as the Scout, Mastiff, and Pioneer. These RPVs, weighing between 165 and 430 pounds at takeoff and powered by two-cylinder 14- to 26-horsepower engines, can be adapted for a variety of missions. Most importantly, they are designed for operation by ground troops with a minimum of training. Besides designing and building RPVs, Israel has tens of thousands of hours of experience in flying RPVs, usually over hostile territory. Incidentally, the U.S. Navy is currently operating Pioneers and the U.S. Marine Corps has Mastiff Mark IIIs.

Israeli Aircraft Industries

Fig. 2-11. The Scout is recovered using a net.

Israel again showed the advantages of RPVs in 1982 when the Syrians deployed SA-6 air defense missiles in the Bekaa Valley in Lebanon. Israel equipped some of its RPVs with special radar deflectors so they appeared as full-sized Israeli fighters on the Syrian radar screens. These overflew the air defense sites, drawing fire from both guns and missiles. This confirmed the fact that the deadly SA-6 had indeed been deployed. While the RPVs drew Syrian fire, manned reconnaissance aircraft photographed the SAM sites and gathered important electronic data on the radio frequencies in relative safety. The RPVs also caused the Syrians to turn on their radars so the Israelis could fire anti-radiation missiles at the sites using the emitted radar signals to home in on. Finally, while the Syrian air defenses were preoccupied with the RPVs, manned fighters went in and destroyed them—without the loss of a single Israeli pilot.

Israel's success with RPVs has been credited with the current high level of interest in UAVs—at least in the United States.

Aquila

Few aircraft programs, manned or unmanned, have brought as much controversy as the Army's Aquila. The Lockheed-developed Aquila was one of the most sophisticated UAV systems ever built. The flying portion of the system consisted of a 260- to 330-pound (depending on the mission) craft. It had an overall length of just under seven feet and a wingspan approaching 13 feet. It was made mostly of a high-tech material, preimpregnated Kevlar-epoxy, giving it stealth characteristics against enemy radar detection.

The Aquila had a top speed of 130 mph and could loiter at around 80 mph for three to ten hours. It could operate at altitudes of nearly 15,000 feet. The two-bladed pusher propeller was powered by a 24-horsepower air-cooled two-cylinder engine. Aquila was designed to carry interchangeable payloads weighing

Lockheed Missiles and Space Company

***Fig. 2-12.** The* Aquila *used a shrouded pusher propeller—both for better efficiency and to prevent entanglement in the recovery net.*

from 80 to 115 pounds located in an underbelly compartment. Typical payloads consisted of a daylight black and white TV camera, laser rangefinder/target designator, or *Forward-Looking Infrared* (FLIR) sensor. Missions planned for the Aquila included target acquisition, laser target designation for laser-guided weapons such as the Copperhead and Hellfire missiles, artillery fire adjustment, and reconnaissance. Later, communications relay, weather reconnaissance, and electronic warfare were added to the list, and were an example of the "requirements growth" that was part of the reason for the Aquila's demise.

The Aquila flew a preprogrammed course with the capability for ground controllers to change the course or command the craft into a loiter mode. Because the Aquila was designed to operate over hostile territory, communication channels for flight commands and sending of TV and sensor pictures were highly jam-resistant.

Lockheed Missiles and Space Company

Fig. 2-13. Like the Scout, the Aquila was catapult-launched.

The Aquila UAV was launched from a hydraulically operated catapult mounted on a five-ton truck. After a mission, the UAV was recovered by flying it into a vertical net mounted on another five-ton truck. The Aquila's pusher propeller was shrouded so it did not get entangled in the net. There was also a parachute as a backup recovery method.

The nerve center for the Aquila system was the *Ground Control Station* (GCS). Here the data from the RPV were processed and displayed, and the UAV flight and mission equipment controlled. Inside this mobile station there were control and display consoles, TV monitors, a computer, and other sundry gear needed to support the Aquila. The station was hardened against nuclear and conventional attack. Two-way communications with the airborne vehicle was via a trailer-mounted antenna. The entire Aquila system was mobile so it could be moved relatively easily to keep pace with changes on a fluid battlefield.

A look at the Aquila on a typical reconnaissance mission gives a good insight into how UAVs can be used in general. The three-man crew sits in the GCS. The mission commander receives mission requirements from the units needing reconnaissance information, then plans the mission and transmits the information back to the people who wanted it. The air vehicle operator is responsible for getting the UAV to and from the target, as well as commanding the vehicle to fly the desired flight profile while the cameras and sensors obtain their pictures. The payload operator's main job is to monitor the pictures being sent back from the UAV; he might, for example, command the TV camera to zoom in to get a more detailed look at the target.

The Aquila program was started in 1974 and was finally killed in 1987, a period of some fourteen years during which it was the subject of considerable debate in Congress and the target of several budget cuts. Along with added-in missions, the cost grew to nearly a million dollars a copy. The number the Army planned to buy dropped from just under 1000 to 548, then to a mere 376. The

Lockheed

Fig. 2-14. *In an inhouse-sponsored program, Lockheed has developed the Aquila into a low-cost Altair UAV to meet new military UAV requirements.*

bottom line was that the Army would have obtained fewer than 40 percent of the Aquilas originally planned, but at about four times the originally projected cost.

While the Aquila was far from a "cheap and simple" UAV, it was a very capable system and still substantially less expensive than manned aircraft for the same job. Unlike other cheaper UAVs that are essentially fair-weather birds, the Aquila could fly in all kinds of weather and was highly survivable against enemy detection. The Aquila did reach quasi-operation status with the Army and achieved a success rate of well over 90 percent in nearly 500 test flights. Much technology was developed and many lessons were learned, however, that will be used in future UAV projects. Indeed, Lockheed has continued the Aquila in a much simplified form, called the Altair.

3

UAVs in a Supporting Role

UNMANNED air vehicles perform the dull, dirty, and dangerous support missions. The list of duties for which UAVs have been proposed—and in many cases, actually built—is quite long. These missions often center around flights behind heavily defended enemy lines, flights into environments contaminated by NBC (*Nuclear, Biological, and Chemical*) weapons, loitering at high altitudes for hours or even days, and the classical "suicide" missions.

Both the American experience with the Firebee in Southeast Asia and Israel's extensive use of a variety of RPVs show that UAVs *do* work—and work well—in combat. Currently, the United States is planning to spend several billion dollars on the development of "non-lethal" UAVs.

The different design approaches taken for support UAVs is equally long. Some designs are quite simple, and not much more than a hobbyist's radio-controlled model airplane fitted with a miniature TV camera. Others are nearly as sophisticated as a manned aircraft. An important class of these support UAVs are the designs that can take off and land vertically and hover while performing their missions. These are the RPHs or remotely piloted helicopters.

Current Plans

The current "Department of Defense Joint UAV Program Master Plan" has neatly categorized future support (or, as the DOD terms them, "non-lethal") UAVs according to the ranges at which they are designed to operate.

The *close-range* systems would have a range of about 30 kilometers (about 20 miles) and would be used primarily to locally support tactical units and small ships. Typical missions include surveillance, target spotting and acquisition, disruption of enemy communications, and deception. One UAV task might be to patrol the perimeter of an airbase to detect possible NBC contamination, to assess runway bomb damage after an attack, and to replace or supplement ground-based radars. Such UAVs could be radio-controlled or even tethered to the ground. The bottom line is that these systems must be very inexpensive so they can be deployed in large quantities.

The *short-range* systems would have a range of 150 to 300 kilometers (about 90 to 180 miles). They would be used for target designation, spotting, and acquisition, as well as relaying communications and detecting NBC contamination. All these would be remotely controlled, and because of added complexity, would be somewhat more expensive than the close-range UAVs. They would also fly at low and medium altitudes for fairly long periods of time.

The *medium-range* systems would perform more ambitious missions, such as post-strike reconnaissance after sorties by manned bombers and fighters. These UAVs would have a maximum range of 700 kilometers (about 400 miles) and would be able to fly at high subsonic speeds. They would be launched from the ground, ships at sea, or a manned aircraft, and would either fly a preprogrammed course or be remotely controlled from a succession of ground stations. Because of the sophistication of their missions, medium-range UAVs would be even more expensive.

Finally, there are the *endurance* systems that could stay aloft as for as long as 36 hours, but their maximum range would be 300 kilometers (180 miles). These would be used for such missions as reconnaissance, communications relay, target acquisition, NBC detection, and weather observation. Like all the systems in the plan, they could be used over land or sea.

Pioneer

Before we look at UAVs of the future, let's look at the UAVs the U.S. military is currently using. These include the Pioneer which is in service with the U.S. Navy and Marine Corps.

The Israeli Pioneer is a direct descendant of the highly successful IAI Scout and Tadiran Mastiff. The Pioneer is a multipurpose RPV. That is, there is a basic system that comprises the 330-pound aircraft with its TV camera and radio-control equipment for guiding and controlling the aircraft, an autopilot to make the aircraft easy to fly, and a power supply for a variety of different mission payloads. The latter includes daylight TV and FLIR sensors, *Electronic Warfare* (EW) and *Electronic CounterMeasures* (ECM) equipment, communications relay gear, and laser equipment for target designation and rangefinding. The Pioneer, which has a 10.8-foot wingspan and an 11.8-foot-long fuselage, is powered by a 26-horsepower, two-cylinder engine driving a pusher propeller.

Ground equipment is also part of the Pioneer package. The ground control station consists of controls, computers, and displays used by the two-man crew, an aircraft operator and observer. The Pioneer is launched using its tricycle gear

USN: photographer: PHC Jeff Hilton/P-13 Striker

Fig. 3-1. A Pioneer I is recovered after a mission in a net erected on the stern of the battleship USS Iowa.

for a conventional (but short-distance) takeoff or it can be catapulted from a rail launcher with either pneumatic (compressed air) or a rocket booster. The Pioneer can be recovered by landing on its wheels with the stopping distance greatly shortened by an arresting cable, or it can be captured in a vertical retrieval net.

The first Pioneers went into service with the U.S. Navy in 1987 onboard the USS *Iowa*, where they are used for gunfire support, *Over-The-Horizon* (OTH) targeting, and reconnaissance out to a range of 110 miles. The Pioneer has an endurance of up to eight hours. Pioneer systems have also been deployed aboard another ship and with the Marine Corps, where they are used in ground-based operations. While the first of these Pioneers were built in Israel, they are now produced by AAI in the U.S. as well.

Amber

The Amber UAV is a good example of a current UAV designed in the United States. The Amber was developed by Leading Systems, Inc., for the *Defense Advanced Research Projects Agency* (DARPA) to carry a variety of payloads including a daylight TV camera, FLIR sensor, communications relay equipment, surveillance radar, COMINT (*COMmunication INTelligence*) or ELINT (*ELectronics INTelligence*) gear. The Amber is a long-duration UAV that can stay aloft for up to weeks at a time. The latest available information is that an Amber has stayed aloft for 38 hours, flying at altitudes up to 27,800 feet, to set a new endurance record. Amber UAVs are also being used to test new payloads, such as moving target indicators using radar and SIGINT (*SIGnal INTelligence*).

DARPA

Fig. 3-2. The Amber UAV is currently in operation with U.S. military services.

Depending on the mission, Amber has an overall length of 13 to 20 feet, a 25- to 39.4-foot wingspan, and weighs 750 to 1500 pounds fully loaded with payloads ranging from 60 to 400 pounds. The structure is made primarily of carbon-epoxy composite and it features retractable tricycle landing gear. Power comes from an 800cc 65-horsepower four-cylinder engine.

The Amber has a top speed of 135 KTAS (153 mph), and it can cruise at 80 KTAS (91 mph) and lands at 60 KTAS (88 mph). While the Amber is now being deployed with U.S. military forces, a second generation Amber II is being developed.

Remotely Piloted Helicopters

Remotely piloted helicopters, or RPHs, combine the advantages of the helicopter (namely, the ability to take off and land on tiny pieces of real estate) with the merits of the UAV. Thus, throughout the history of UAVs, starting with the Gyrodyne QH-50, RPHs have been proposed, designed, built, and test-flown for a variety of missions. Among the earliest were a whole series of British-designed RPHs

Fig. 3-3. The Westland Wideye was a British-built RPH of the late 1970s. The twin counter-rotating rotors eliminated the need for a conventional tail rotor.

Westland Helicopters Ltd.

Fig. 3-4. Another Westland RPH prototype of the 1970s was the Wisp.

with counter-rotating rotors; these reached the prototype stage. Westland Helicopters built RPHs named Mote, Pupil, Supervisor, Wideye, and Wisp. These were succeeded by the ML Aviation Sprite of quite similar design in the 1980s.

Canada's Canadair, Inc., has developed several UAVs including the very interesting CL-227 Sentinel, which has been nicknamed "Peanut" because of its hourglass shape. The shape was chosen to reduce the UAV's radar detectability to a minimum. For the same reason, the CL-227's body and rotors are made of radar-absorbing materials such as Kevlar and the engine inlet and rotor downwash are designed to reduce infrared detectability. While early versions were powered by a Wankel rotary engine, current CL-227s use a small turboshaft. The engine is housed in the upper module while the payload and communications are housed in the lower module. The counter-rotating rotors are located between the two. The payload module is easy to change so the Sentinel can be quickly adapted for a variety of missions.

The CL-227 has a top speed of over 80 mph—and, of course, it can hover. It can stay aloft for up to four hours. With a maximum takeoff weight of over 400

ML Aviation Ltd.

Fig. 3-5. The ML Aviation Sprite, with its round fuselage and four landing legs, might be mistaken for a UFO.

pounds, it can carry up to a 100-pound payload. The Sentinel can be radio-controlled with sensor data being transmitted to the ground by a radio link, or it can be tethered to the ground (or, as usually proposed, to a ship).

The *Airborne Remotely Operated Device*, or AROD, is a rather simple UAV developed by the Sandia National Laboratories for reconnaissance and surveillance missions over enemy lines. Powered by a small piston engine, the 85-pound AROD carries enough fuel to stay aloft for about an hour. It can be controlled either by radio control or while tethered via a fiber-optic cable. While the latter limits the AROD's range, it does prevent jamming of the radio frequency link between the ground and the airborne vehicle.

Stabilization of the AROD is done by an onboard computer and the AROD's direction of flight is controlled by four control vanes. A portable ground control unit houses the electronics, TV monitors, and joysticks for controlling both the craft and the onboard cameras.

While the demonstrator unit carries two high-resolution black-and-white TV cameras operated from the ground, other types of sensors could also be fitted.

APPLICATIONS

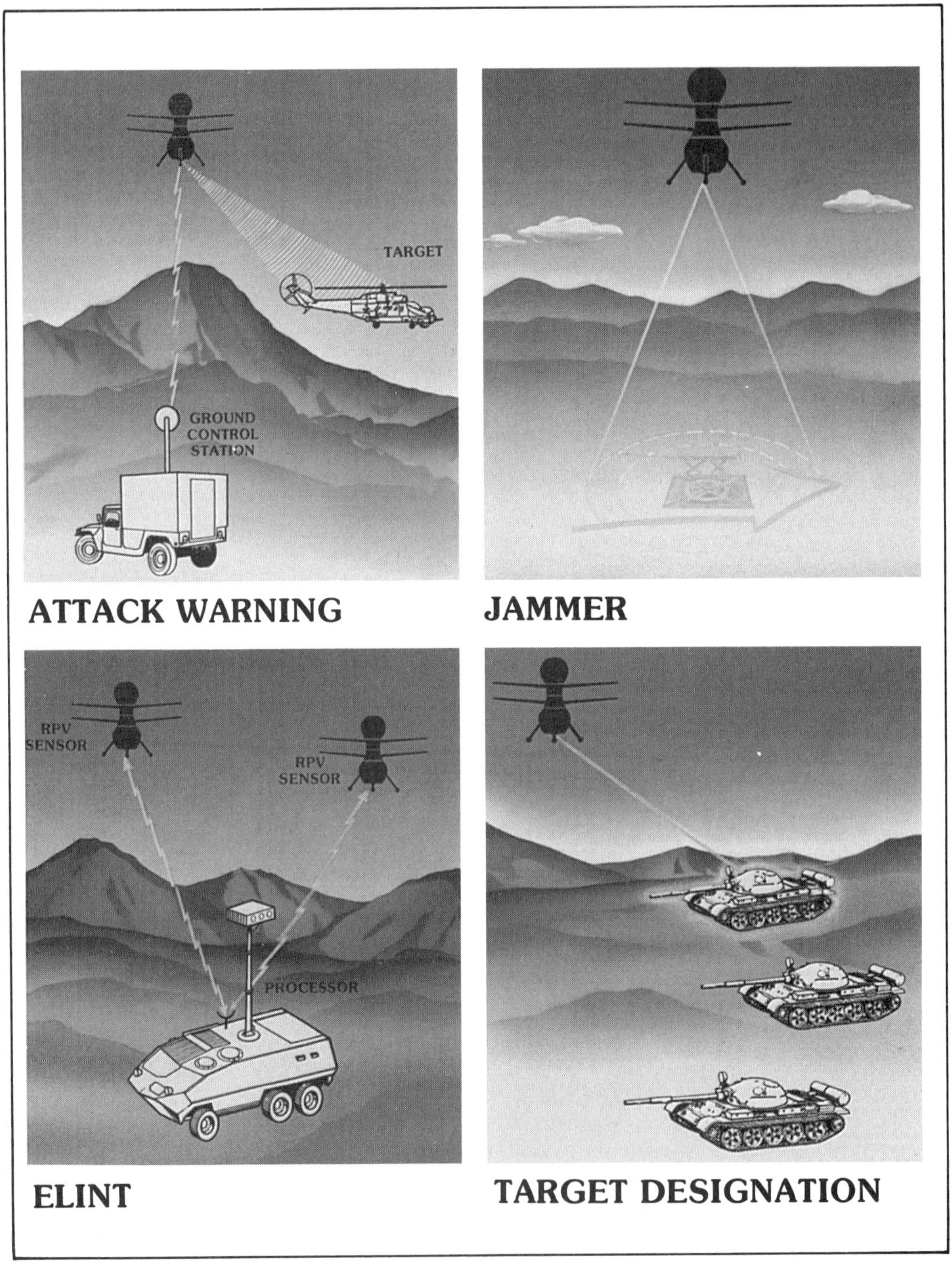

Fig. 3-6. Some applications for the Canadair CL-227. Other missions could be as a decoy, for ship-to-shore bombardment, over-the-horizon surveillance, and sonobuoy deployment.

Sandia National Laboratories

Fig. 3-7. The AROD could be used to gather intelligence on the battlefield.

Even though the unit was designed for military applications, it could be easily adapted for civilian surveillance jobs such as watching for intruders at a secure facility.

The Pointer is an interesting UAV concept that combines the best features of fixed-wing and rotary-wing aircraft. The Pointer is really a miniature version of the V-22 Osprey *Vertical/Short TakeOff and Landing* (V/STOL) manned aircraft being developed by Bell Helicopters and Boeing Vertol for the U.S. military services. With its tilt rotor, the Pointer can fly at fast forward speeds while still being able to take off and land vertically and hover. High speeds are desirable to get the UAV to and from the mission quickly. Helicopter-like flight eliminates the need for complex recovery systems, launching catapults, and recovery nets. Also, the landings are gentler—especially to sensitive electronics—and thus repair and refurbishment after each flight is greatly reduced.

The Pointer is much smaller than the manned Osprey, which is designed to carry troops (among other things). The Pointer has a wingspan of under 11 feet, a

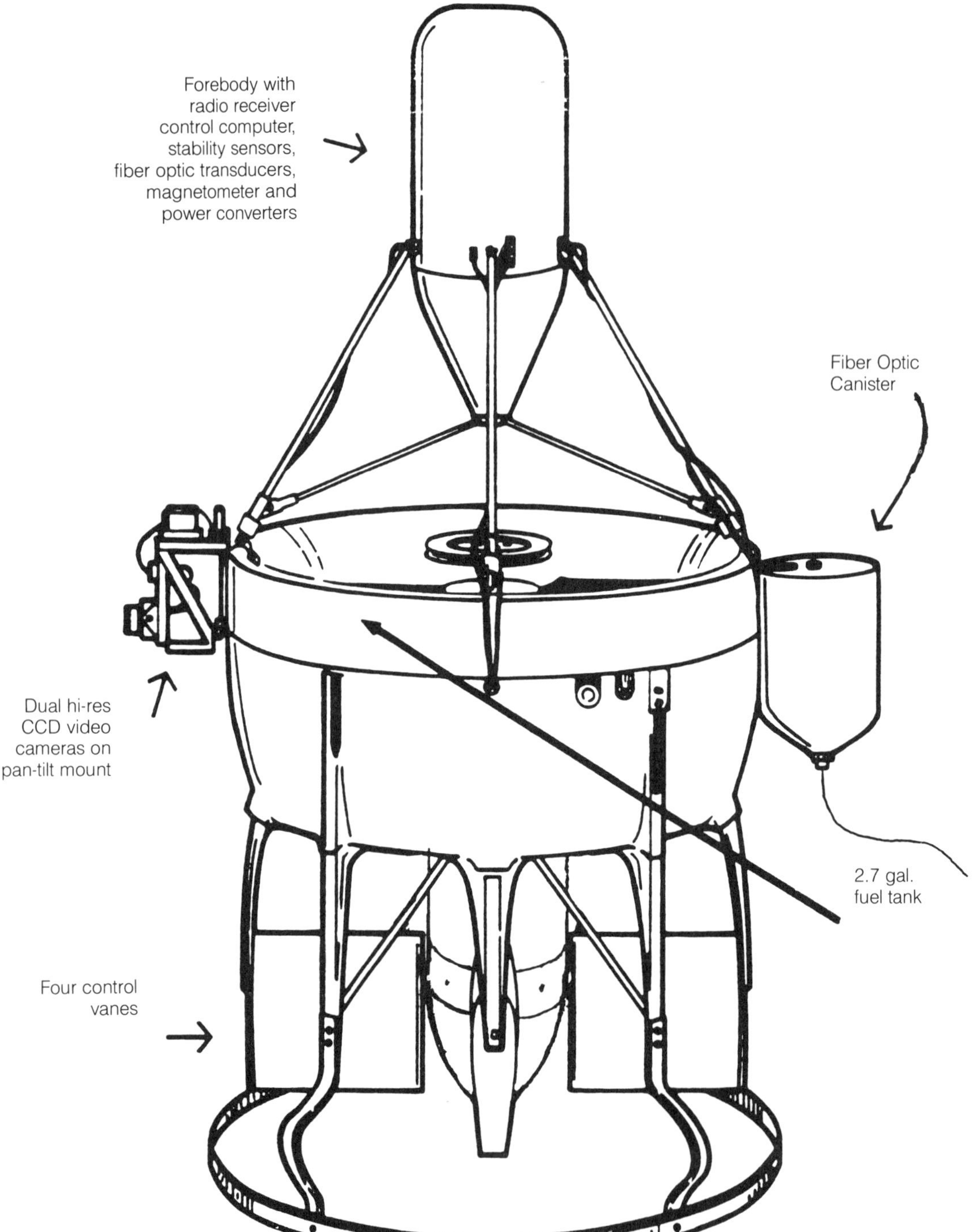

Fig. 3-8. Layout of the Airborne Remote Operated Device (AROD). Sandia National Laboratories

Boeing Helicopters

Fig. 3-9. The Boeing/Bell Pointer is the world's first tilt-rotor UAV. It made its first flight in 1988.

fuselage length of just over 13 feet, and a total weight of just over 600 pounds—about one percent of the weight of the Osprey. Power will come from a gas turbine engine that produces around 100 horsepower, although the prototype uses a 95-horsepower snowmobile engine for flight testing.

Speaking of flight tests, the control system for the prototype consisted of "off-the-shelf" RC equipment familiar to any radio-controlled model aircraft enthusiast. Some commercial RC equipment is now so good that it came to be used in serious UAV development programs. Testing also pointed up another important fact about flying UAVs: Because things happen so rapidly with these small-scale craft, many pilots have difficulty flying them. However, for the experienced radio-controlled model aircraft "pilot," it is a piece of cake!

4

Weapons and Targets

DESTROYING enemy targets was one of the first uses envisioned for unmanned aircraft, starting with the concepts of "Boss" Kettering and Professor Low during World War I. In World War II, remotely controlled weapons such as the American Bat and German Ruhrstahl SD. 1400X saw limited action. Smart weapons such as the Maverick missile and TV-guided bombs proved their worth in the Vietnam War. The dividing line between guided missiles and UAVs as weapons often becomes a bit blurred; thus, weapons such as the highly successful wire-guided TOW and laser-guided Hellfire missiles or the GBU-15 TV-guided glide-bomb could logically fall within the definition of the UAV, since they are all remotely controlled.

Using unmanned aircraft for target practice is an idea almost as old as using them for weapons. Starting with Tiger Moths, obsolete aircraft have been fitted with radio-control equipment so they could be shot at by antiaircraft gun and surface-to-air missile crews, or by pilots in other aircraft. There have also been a multitude of RPVs that have been developed and produced *strictly* as targets.

Anti-Radar for Defense Suppression

The UAV is an ideal weapon for homing in on and destroying enemy radar installations so manned aircraft can fly to their designated targets unhindered by enemy air defenses. The Israelis dramatically demonstrated this tactic in its "Peace for Galilee" offensive against Syrian air defense installations in the Bekaa Valley. The United States already has the Harm anti-radar missile in its inventory. These anti-radar weapons have one thing in common: They home in on radar emissions. Logically, these weapons are also called *anti-radiation weapons*. The enemy radar therefore has a choice of shutting down or being destroyed; in either case, the radar is rendered impotent.

One of the next-generation defense supression weapons is the AGM-136A Tacit Rainbow being developed jointly for the Air Force and Navy by the Northrop Corporation. While Tacit Rainbow is designed to be air-launched, it could also be launched from the ground; thus the Army is also interested in the weapon. After launch, the AGM-136A flies to the target area, where it can loiter for long periods of time while searching out signals emitted from enemy radar installations; the latter can be either on land or on warships at sea. When a signal is detected, the missile automatically homes in on the radar beam and attacks the radar site. If radars are turned off, the Tacit Rainbow will persistently loiter in the area until they are turned on. Then the missile will detect and destroy the source of the signal.

Northrop Corp.

Fig. 4-1. *The Tacit Rainbow anti-radar weapon.*

Northrop Corporation

Fig. 4-2. *Tacit Rainbow being launched, from A-6 Intruder during the flight test program.*

The Tacit Rainbow is a "fire-and-forget" weapon; once it is launched, it flies autonomously without further outside commands using its onboard computer and advanced avionics. To allow more missiles to be carried on the launching aircraft, the Tacit Rainbow's wings and elevons are folded before use and then unfolded prior to launch. To keep costs down, the AGM-136A uses many plastic and aluminum parts that are bonded or bolted together rather than joined by more expensive riveted or welded construction. It is powered by an air-breathing jet engine. One launch vehicle being considered for the AGM-136A is the venerable B-52.

Another anti-radar weapon being considered by the Air Force is the Seek Spinner anti-radar attack drone. This UAV is based on the BRAVE-200 (for *Boeing Robotic Air VEhicle*) being developed by the Boeing Military Aircraft Company. The BRAVE-200, which was begun in 1979, is an ultra-low-cost multipurpose robotic air vehicle. Different missions can be performed by changing the payload in the nose of the seven-foot-long multipurpose airframe.

Besides being used as an anti-radar weapon, the Seek Spinner could perform other missions such as electronic warfare, battlefield surveillance, search-and-destroy missions deep behind enemy lines, and relaying of communications. For instance, in the Pave Cricket program, the Air Force plans to demonstrate that the BRAVE-200 can be used as a low-cost means to jam enemy battlefield communications and sensors. In each application, a common BRAVE-200 airframe would be used, with the modular payload being changed. By using a common airframe, production costs could be greatly reduced through economies of large production runs. To further reduce costs, the BRAVE's basic structure is

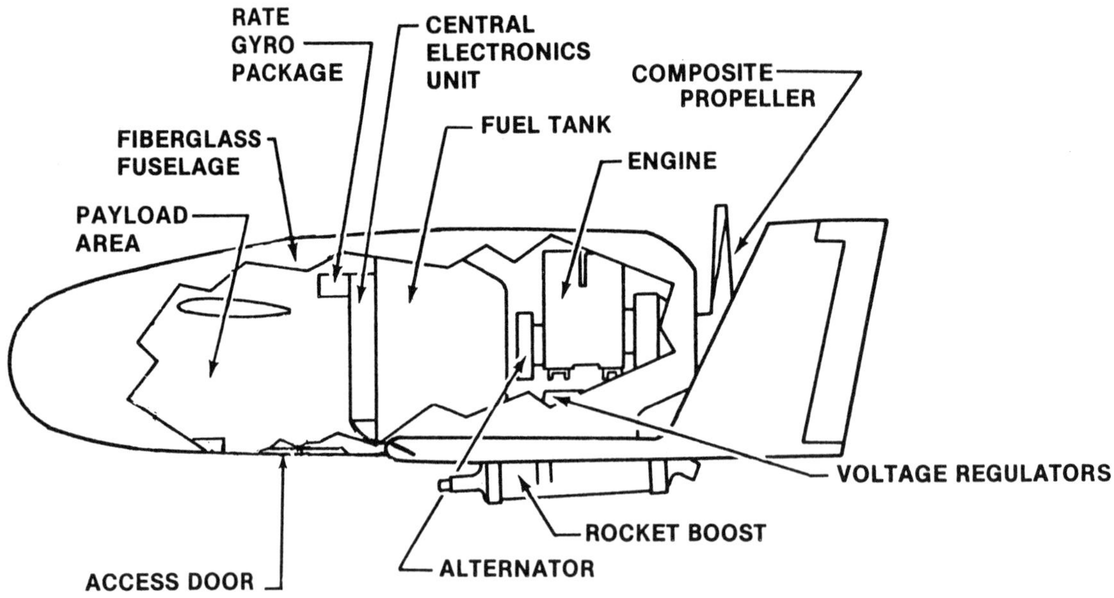

Fig. 4-3. Layout of components of the BRAVE-200. Boeing Military Aircraft Company

made of durable injection-molded composite materials. This not only means low cost, but allows the BRAVE to be produced in large quantities. The 265-pound BRAVE-200 is powered by a 28-horsepower 438cc two-cylinder engine driving a four-bladed pusher propeller.

One of the biggest advantages of the BRAVE-200 is that it is a "fire-and-forget" vehicle. The basic system has a built-in dead reckoning navigation system. Once in flight, the vehicle's canard/elevator, spoilers, and rudders are steered by a preprogrammed flight control system that includes a digital autopilot. Because

Boeing Military Aircraft Company

Fig. 4-4. The BRAVE-200 in flight.

of the BRAVE-200's modular design, other navigation systems could be incorporated for other missions or greater accuracy. As another cost-saving consideration, the fuel tank is located at the BRAVE's center of gravity so aerodynamic stability can be maintained automatically as fuel is consumed.

The BRAVE-200 can be serviced, programmed, and launched by a two-person crew. Because it is rocket-launched, it requires minimum real estate and essentially no launch site preparation. The BRAVE-200 is designed for long-term storage in modular containers that can hold batches of 2, 4, 6, 9, 12, or 15 of the BRAVE-200 systems. Depending on the number of units, the containers can be carried on trucks, trains, or ships. For example, the 15-vehicle container is the size of the standard 8-×-8-×-20-foot international shipping container. The BRAVE-200 can be stored in the sealed containers for years. Current specifications call for five-year storage with a 15-year goal—and without maintenance. When the BRAVEs are required, the container would be powered up and the drones would be fueled. Then each vehicle would emerge as its 8.5-foot span wings are unfolded into their swept position for flight. (The wings are folded along the fuselage for denser storage.) The BRAVE-200 would then be elevated

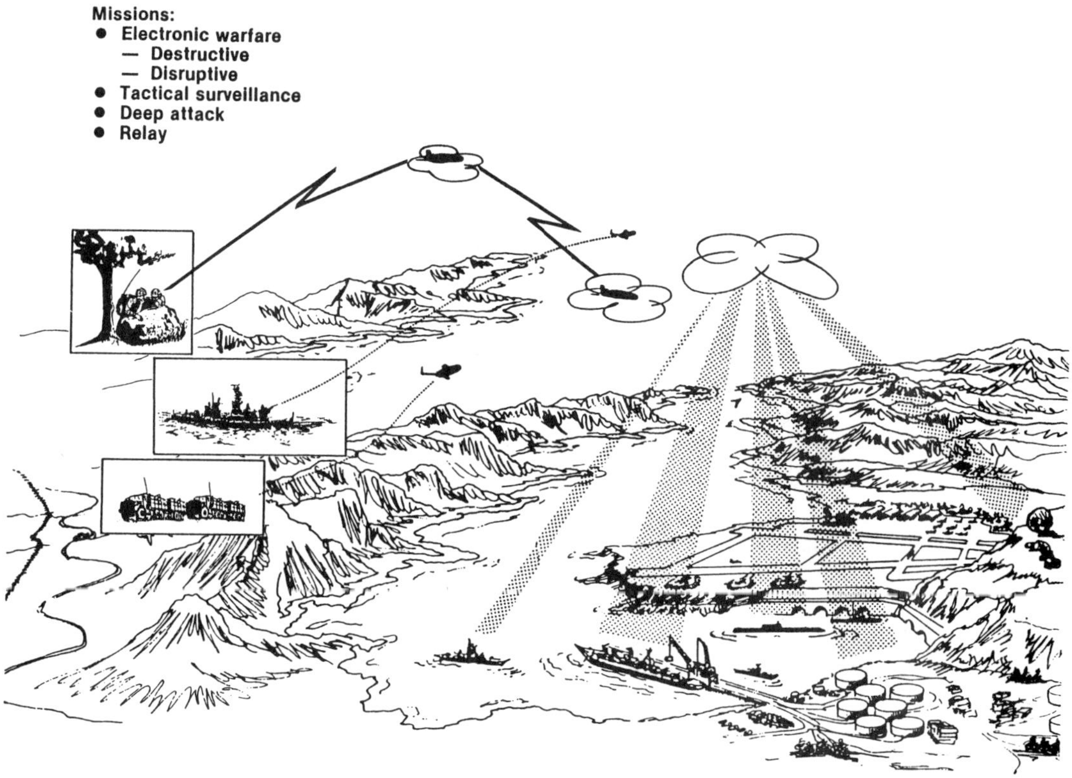

Fig. 4-5. Besides use as an anti-radar weapon, the BRAVE-200 could be used for a variety of lethal and non-lethal missions. Boeing Military Aircraft Company

to a 20-degree launch attitude and the starter motor would engage the propeller hub to start the engine. After an automatic go/no-go checkout and programming of the guidance system, the BRAVE-200 would be launched with a rocket bottle to provide additional initial thrust.

On a typical mission, the BRAVE-200 would climb to an altitude of 8000 to 12,000 feet enroute to the target. There it could attack immediately, or reduce its speed and loiter for jamming or surveillance missions. Later, it could move to another area to complete another assigned mission. Of course, this would all be programmed before launch. The BRAVE-200 has a cruise speed of 140 mph and a loiter speed of 90 mph.

Besides the BRAVE-200, Boeing has developed the jet-powered BRAVE-3000, which would also be rocket launched. The BRAVE-3000 could cruise at 400 mph. It also would be expendable. Low-cost, autonomous, expendable UAVs could overcome the problems of cost growth and oversophistication that have killed RPV programs in the past.

The United States is not the only country interested in defense suppression UAVs. Dornier in West Germany has pursued development of anti-radar drones such as the *KleinDrohne Anti-Radar* (KDAR), Hornisse, and *Drohne Anti-Radar* (DAR). These are all mini-drones that are ground-launched. For example, the DAR is about 7.5 feet long with a 6.5-foot wingspan; it weighs less than 250 pounds. Power comes from a 26-horsepower two-cylinder two-stroke engine. Messerschmitt-Bolkow-Blohm has also pursued its own version of the DAR, which is about the same size as the Dornier DAR. Israel, the nation with the most actual experience in using defense suppression techniques under combat conditions, is developing UAVs such as the Harpy.

Weapon Platforms

While the UAV was for decades touted as a natural low-risk bombing platform, most proposed unmanned weapon platforms such as the Gyrodyne QH-50 and Ryan Firebee discussed earlier did not get much beyond the prototype stage. Again, to be completely correct, there is a proliferation of lethal UAVs if you include missiles within the definition. Viewed in this context, lethal UAVs are already highly refined and successful.

Martin Marietta Electronic Systems is pursuing what might be the ultimate in UAV technology with its *Autonomous Air Vehicle* (AAV). The AAV is designed to fly through enemy lines carrying several self-guided submunitions. Its onboard computer and sensors, such as millimeter wave radar and forward-looking infrared (IR) detectors, would locate targets on the ground even if well-camouflaged. Millimeter radar is able to see through rain, fog, and snow.

The AAV's *Geometric Arithmetic Parallel Processor* (GAPP) computer allows the system to not only "see" the target, but, most importantly, determine that it is the *right* target. The GAPP uses a parallel-processing neural network computer that mimics the functioning of the brain to handle the volume and complexity of computations needed to "understand" what the sensors are viewing.

When the proper target is identified, the AAV would launch its weapons and then stay around to assess the damage. If need be, it would strike again. The AAV could also include the ability to avoid enemy aircraft and missiles that could destroy it. Finally, it would return home to be recovered and used again. While the AAV will probably never go into production, the advanced technology being developed could be used in all types of "very smart" weapons including missiles and munitions.

Unmanned Targets

Remotely piloted technologies have already proven their worth many times over when it comes to being used as targets for maintaining the proficiency of weapons crews as well as in evaluating the effectiveness of new weapon systems.

Aircraft such as the B-17, B-47, F-80, T-33, T-38, F-86, F-100, F-102, F-104, and F-4 have been converted into *Full-Scale Aerial Targets* (FSAT). These obsolete or worn-out aircraft, which normally would have gone to the scrap yard, are given a "Q" designation (i.e., QF-86, QF-4, etc.) and a new lease on life. Typically, the target drones are modified with radio-control equipment, autopilots, and other electronic systems for autonomous flying, as well as electronic gear for scoring and perhaps TV systems to help in takeoffs and landings.

As an aside, in some cases manned aircraft have been converted to unmanned aircraft for other than target purposes. For example, Sperry Defense Systems converted several Beechcraft Bonanza utility aircraft into QU-22 drones. These were used as airborne relay stations to transmit data on enemy infiltration obtained from acoustic sensors installed on the South Vietnamese side of the DeMilitarized Zone (DMZ). While they flew pilotless during missions that could last as long as 18 hours, a human pilot was still onboard to monitor the operation and for takeoffs and landings. Reportedly, the Soviet Union has used aircraft such as the Yak-25 Mandrake converted to radio-control operation for unmanned reconnaissance and electronic intelligence work.

Nowhere is there a greater variety of UAVs than in the craft designed specifically for target work. Some are large, complex, and highly sophisticated; others are no more than R/C model aircraft much like those you might purchase at a hobby shop. Indeed, few aircraft of any type have been built in such large quantities or over such a long duration as the Australian Jindivik and French Aerospatiale CT.20, and the numerous target drone models from Beechcraft, Northrop, RS Systems, and Teledyne Ryan in the United States. Besides countries like the U.S., Britain, Germany, and France, target drones are built and used in countries with more modest aerospace forces and industries such as Argentina, Brazil, and Switzerland.

The diversity of targets simulated by these drones is as diverse as their designs. Several manufacturers make target models of threat aircraft to train anti-aircraft and surface-to-air missile crews. Soviet aircraft such as Su-25 Frogfoot and MiG-25 Flogger-D fighters and the Hind-D attack helicopter are popular. They come in 1/5-, 1/7-, and 1/9-scale versions, for example. Powerplants include typical model airplane glow-plug engines; the larger models use engines adapted from snowmobiles, chainsaws, ultralight aircraft, and even motorcycles. Radio equipment can be as simple as four- or six-channel proportional controls and transmitters that are used by R/C model enthusiasts. "Off-the-shelf" components

Aerospatiale

Fig. 4-6. After its mission is over, a French C 22 Target Drone is recovered via a parachute.

Farner Air Service

Fig. 4-7. The Swiss-built Farner KZD-85 Topaz is used for antiaircraft gunnery and missile training with the Swiss Army. It is powered by a 99cc, 6.5-horsepower engine.

Fig 4-8. The controls for the Farner Topaz would be familiar to any R/C model enthusiast.

Farner Air Service

Target Technology, Ltd.

Fig. 4-9. The BTT-9 Frogfoot is an R/C model of the Soviet Su-25 ground attack aircraft. It comes in several sizes, with a wingspan of 85 to 109 inches, three optional powerplants (15cc, 32cc, or 100cc) and a top speed of 98 to 140 mph.

from such familiar names as Fox, Futaba, K&B, Tartan, and Kraft are often used.

While smaller than the target they are simulating, these models can be equipped with electronic equipment that makes them look larger—at least on a radar screen. Other times, the scale models are flown at scale distance so they appear "full size" to the observers. Sometimes the target drones are used to pull banners and streamers that are fired at so the drone can be recovered and used again. They can be used as targets for sophisticated weapons such as the Redeye and Chaparral missiles or for small arms and light-caliber machine gun fire. Often these RPVs carry payloads such as gas or smoke grenades or pyrotechnics to add authenticity when they are hit, infrared sources for use with infrared homing weapons, and even cameras to be used in scoring. These UAVs are usually either launched from a rail or hand-launched. The whole idea is to keep them simple and cheap while still providing realistic training.

The *Supersonic Low-Altitude Target* (SLAT) is an example of a more sophisticated "training aide." SLAT is being designed by Martin Marietta for the Navy to simulate anti-shipping missiles that travel at sea-skimming altitudes. It will be

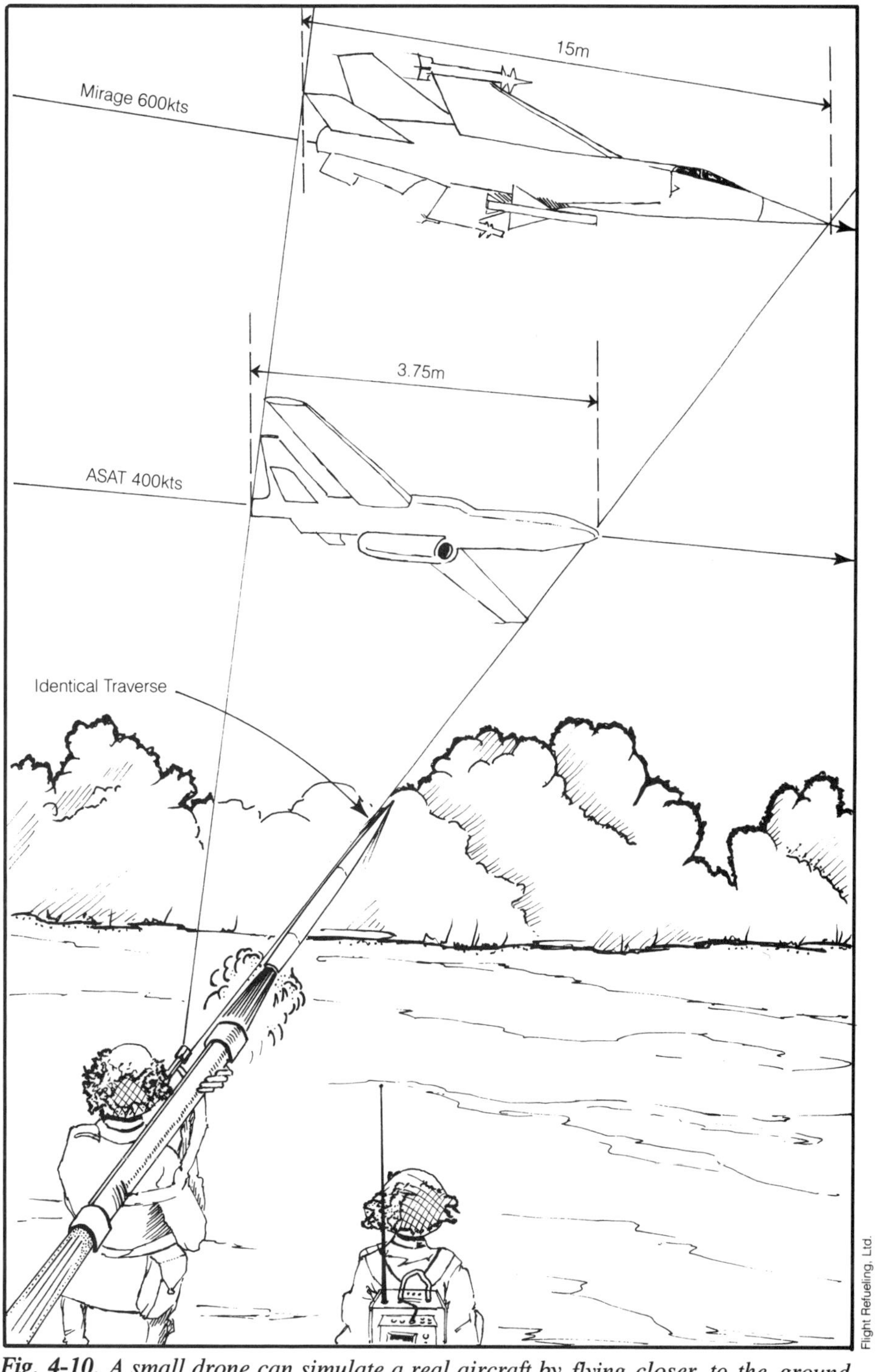

Fig. 4-10. A small drone can simulate a real aircraft by flying closer to the ground and at a slower speed.

Martin Marietta

Fig. 4-11. The Supersonic Low-Altitude Target (SLAT) would be used to train shipboard anti-shipping missile crews.

used to test the effectiveness of crews and equipment in countering low-level cruise missile attacks.

After air-launch at an altitude of about 1000 feet and about 60 miles from the ship, the SLAT drops to an altitude of a mere 30 feet and travels at Mach 2.5. During its flight, the SLAT duplicates the maneuvers of an actual cruise missile. SLAT is propelled by a integral rocket/ramjet propulsion system, which allows greater range in a compact package. The solid rocket portion of the engine is used for the initial boost and then the spent rocket case becomes the combustion chamber for the ramjet. After its simulation mission is over, the SLAT would be recovered using a parachute with retrieval accomplished by either helicopter or boat. SLAT is designed so it can be recovered, refurbished, and reused at least four times.

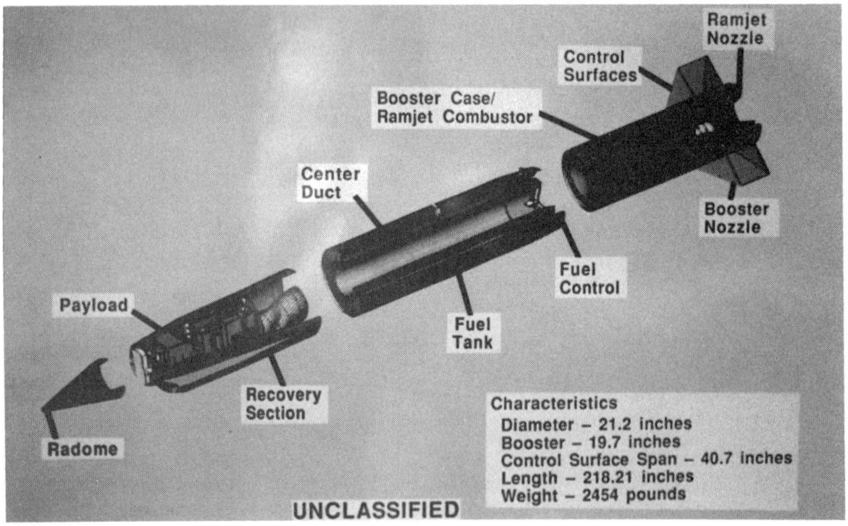

Fig. 4-12. The SLAT's spent solid propellent booster rocket case becomes the combustion chamber for the ramjet engine.

5

Remotely Piloted Research Vehicles

AEROSPACE engineers have often used unmanned air vehicles to flight-test new ideas. A special term has even been coined for these test aircraft: *Remotely Piloted Research Vehicles*, or RPRVs. RPRVs have been built in a variety of forms, from rather simple radio-control models of aircraft that can take off and land vertically to sophisticated large-scale models of military fighters for the 21st century.

Why RPRVs?

RPRVs have added a new dimension in the development of experimental aircraft. Before RPRVs, if engineers wanted to try out a new aircraft concept, they tested it in a wind tunnel, relied on computer calculations, or built and flew a full-scale prototype. Wind tunnel testing and computer calculations do not usually provide *all* the information needed to prove that a new aircraft design will fly as designed—and do it safely. Building a prototype to be flown by a human test pilot is both very expensive and time-consuming.

The RPRV provides an intermediate means to provide "proof-of-concept" information quite inexpensively. Much of the cost savings come from having the sophisticated electronics, computers, and flight controls on the ground rather than in the air. The RPRV can get by with simpler (and thus less expensive) onboard equipment. Much of the equipment does not have to be miniaturized and hardened for the rigors of flight because it remains on the ground, greatly reducing costs. If the RPRV crashes, the expensive equipment is not lost. A reduction

Lockheed-Georgia Co.

Fig. 5-1. Remotely Piloted Research Vehicles (RPRVs) provide an inexpensive and safe means to try out new aerodynamic concepts. This triple-engine, 1/3-scale model is flown like any R/C model.

in onboard equipment leads to smaller, lighter-weight, and thus less costly airborne test platforms. Finally, an RPRV can be built more economically because it does not have to be full-scale nor need to be man-rated. Being unmanned, the RPRV does not have life support and escape systems, which are very expensive items.

RPRVs can drastically cut the time from the conception of an idea to actual flight test so engineers can see their ideas at work sooner. Aerospace companies have used RPRVs to demonstrate new concepts to customers to determine if a project is worth pursuing further. In some cases, RPRVs have even been built primarily as sales tools for new ideas, and not as research tools.

RPRVs can be used to identify problems with a particular idea before vast sums of money are spent on a manned prototype. It is much easier and cheaper to modify—or even rebuild—a remotely piloted aircraft that does not work right than to make the same repairs on a manned aircraft. The more daring ideas can be tried out without risking the life of a human test pilot.

RPRVs can speed up and reduce the cost of developing a new aircraft as well as the risk. They can identify early concepts that will not work and help eliminate the bugs from concepts that will work. However, they do not obviate the need to flight-test a manned prototype before a design is ready for production. Like the wind tunnel and computer, the RPRV is another tool in the engineer's and designer's tool chest.

Simple RPRVs can be carried aloft and released from a larger manned aircraft. For example, the venerable B-52 has been the host aircraft for several RPRVs. When a flight test is over, a parachute can be popped and the unmanned aircraft can gently float to Earth or can be snagged in midair by a helicopter to be recovered and reused. RPRVs can be flown using an onboard preprogrammed

Julian Wolkovitch

Fig. 5-2. This radio-controlled model was used to flight-test another unique aerodynamic concept, the joined wing.

computer. The flight of an RPRV can also be controlled from the ground, from the launch aircraft, or from a chase airplane. The most sophisticated RPRVs take off, fly, and land under control of a human "pilot" just like an R/C model aircraft. In many cases, an onboard TV camera can provide the view from the "cockpit."

RPRV Examples

Several years ago, NASA used a modified Firebee II drone to test new aerodynamic concepts, including a device to suppress flutter. While it is desirable to make wings as thin as possible to reduce weight and conserve energy, thin wings can bend—or *flutter*—at high speeds. The RPRV was an economical way to determine if the NASA-designed device located on the trailing edges of the wing would damp out the flutter. In this case, the Firebee was dropped from a B-52 on flights that lasted up to an hour. While the RPRV was controlled from the ground, if radio contact were lost, an autopilot would take over until a human pilot in a chase F-104 could assume control. A parachute snatched by a helicopter was used to recover the Firebee after the flight.

During the development of the F-15 Eagle, the Air Force used a 3/8-scale RPRV to test the aircraft's stall and post-stall characteristics—one of the most

Fig. 5-3. Grumman's concept for its "Nutcracker" Vertical/Takeoff and Landing aircraft was test flown with this unmanned radio-controlled scale-model.

demanding and dangerous portions of the test program for a new aircraft. The 24-foot-long F-15 model was dropped from a B-52 and then the ground-based pilot flew it through all the desired maneuvers. A similar RPRV was used during the development of the USAF's F-16 Fighting Falcon.

One of the more successful and demanding RPRV projects was the joint NASA and Air Force HiMAT (*Highly Maneuverable Aircraft Technology*) pro-

USAF

Fig. 5-4. After being airdropped, this Firebee BQM-34 drone flight-tested new aerodynamic concepts.

gram. The object of this program was to develop and test advanced materials, flight controls, and aerodynamics for highly maneuverable combat fighters. The HiMAT aircraft was about half the size of a real manned fighter and weighed about 3000 pounds. It was powered by a jet engine similar to the one used in the Air Force's T-38 Talon trainer. Like so many other RPRVs, the HiMAT was carried aloft by a B-52. The HiMAT's mission started at 45,000 feet and included maneuvers that resulted in sustained 6- to 8G turns at speeds of 600 to 800 mph. Each flight ended with the HiMAT landing on skid-type landing gear on the dry lake bed at Edwards AFB, California.

The HiMAT's pilot had all the normal controls—stick, rudder pedals, instrumentation, throttle—plus a television screen that gave him essentially the same

NASA

Fig. 5-5. Several RPRVs, such as this 3/8-scale F-15, have been carried aloft under the wing of a B-52.

McDonnell Douglas

Fig. 5-6. The 3/8-scale RPRV of the F-15.

view of the flight as if he were actually in the cockpit. Sensors and a TV camera transmitted information to a ground-based computer that digested the information and sent the results to the pilot. The pilot's control inputs were immediately returned again via the computer and telemetry. The tricky part of the whole operation was the fact that all this had to be done *in real time* because at the speeds the HiMAT was flying, a split-second delay could have spelled disaster. Like most research RPVs, the HiMAT had telemetry systems to monitor the flight and report on how well the craft was working. Typically, this includes such information as engine rpm, temperature, altitude, airspeed, bank and roll, angle of attack, sideslip, and roll. In case the ground-based pilot lost contact, control of the HiMAT could be taken over by another pilot in an F-104 chase plane.

The HiMAT used advanced digital "fly-by-wire" controls, in which hydraulic actuators moved control surfaces in response to electrical signals sent over wires from a central control station aboard the RPRV. Fly-by-wire control systems are common in today's high-performance aircraft; they are sophisticated versions of the simple servos used in R/C models, and thus quite appropriate for use in UAVs and RPRVs.

By using a modular design for the HiMAT's components, the UAV could be rapidly modified to flight-test advanced concepts such as *Controlled Configuration Vehicle* (CCV), supercritical wings, two-dimensional vectored-thrust exhaust nozzle, integrated propulsion control, variable camber wings, and deformable/self-trimming outer wings. Between 1981 and 1983, the two HiMAT RPRVs

Canadair, Inc.

The CL-289 surveillance drone uses a solid rocket booster and a turbojet for sustained flights at up to high subsonic speeds. The CL-289 was developed jointly by Canadair and Dornier GmbH in Germany.

Teledyne Ryan International

The Teledyne Ryan ''Integrated Mobile RPV/LRV Reconnaissance'' is launched from a mobile rail launcher. The Command Control Ground Station is shown on the right.

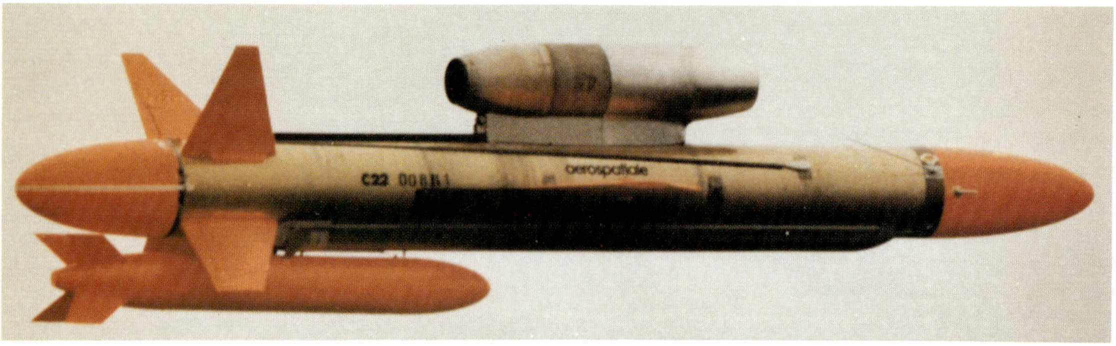

Aerospatiale

The Aerospatiale C 22 target drone is used for air-to-air and ground-to-air firing practice. Propelled by a turbojet engine, it can reach speeds up to Mach .9 and altitudes of 43,000 feet.

U.S. Navy/PHC Jeff Hilton

A U.S. Navy Pioneer I ready for takeoff from the deck of the USS Iowa.

Federal Aviation Administration

The world's largest UAV was this radio-controlled Boeing 720, which was crashed on the dry lake bed at Edwards AFB for research. *(See Chapter 5.)*

Canadair, Inc.

A Canadair CL-227 Sentinel shown in the tethered mode for operation off the deck of a ship. *(See Chapter 3.)*

Developmental Sciences

The Developmental Sciences R4E Skyeye is a rather successful family of UAVs. The latest R4E-50 has gone into service with the Egyptian Army and Air Force and the R4E-30 (shown here) was supplied to the Royal Thai Air Force in the early 1980s.

Beech Aircraft

The Beech MQM-107, a target drone with high subsonic speed, is used by the U.S. Air Force and Army, plus several foreign nations. *(See Chapter 2.)*

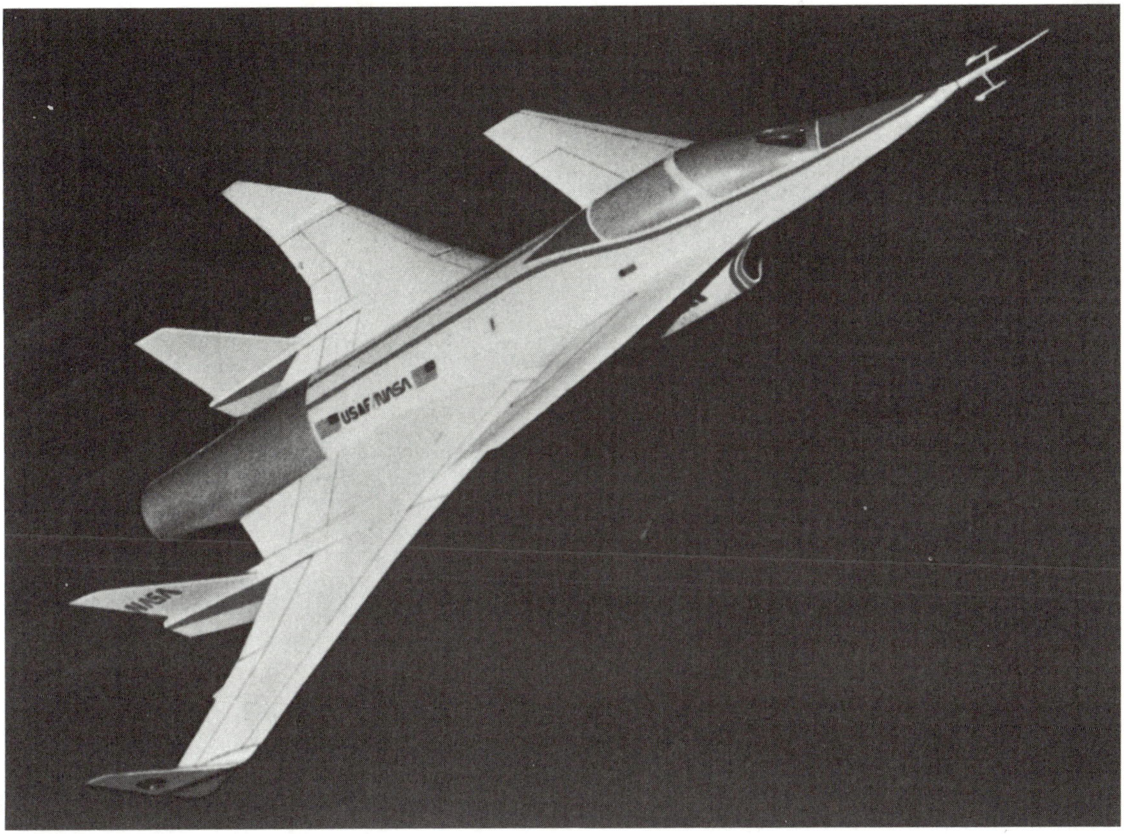

NASA

Fig. 5-7. Two HiMAT RPRVs were built by Rockwell International to investigate techniques that could be used in future highly maneuverable fighter aircraft.

built by Rockwell International made 26 flights. Prior to the HiMAT vehicles themselves, NASA performed flight-testing control techniques for highly maneuverable aircraft, using first an unpowered $^{3}/_{8}$-scale fiberglass F-15 RPRV, followed by two Firebee IIs modified for RPRV duty.

Another example RPRV use to demonstrate unique aerodynamic concepts was the "scissor wing" demonstrator flown in the late 1970s. In its search for economical ways to fly at supersonic speeds without the objectionable sonic boom problem, NASA's premier aerodynamist, Bob Jones (who is credited with inventing the swept wing) came upon the idea of the "scissor" or *oblique* wing. The scissor wing would have a thin airfoil that would require less power for takeoffs, landings, and low-speed flight. For high-speed flights (around Mach 1), the entire wing would be pivoted as much as 60 degrees. This would accomplish the same thing as a variable-swept wing, except it could be simpler because the single span could be pivoted without resorting to a complicated and heavy hinge-like mechanism.

Naturally, building a full-scale aircraft to test something as radical as the

Fig. 5-8. This RPRV was used to test the scissor wing concept.

Fig. 5-9. Two views of the oblique or scissor wing RPRV in flight. On the left, the wing is pivoted to 15 degrees; on the right, the angle has been increased to 30 degrees.

Fig. 5-10. The RPRV pilot flies the craft from this control console. The TV monitor shows the actual view out of the airplane.

NASA

scissor wing would be very expensive—as well as a bit risky! Thus, an RPRV version was built and tested. The scissor wing RPRV was a 900-pound, 25-foot-long aircraft made mainly of fiberglass and powered by a 90-horsepower air-cooled engine. While a real scissor-wing aircraft would fly at speeds around 750 mph, the engineers still learned a great deal from the RPRV version (which had a top speed of under 100 mph). This RPRV took off from the ground under remote control and landed in the same manner. The ground-based pilot flew the craft using both TV pictures and telemetry information transmitted from the airborne aircraft to his control station.

To date, the biggest remotely controlled aircraft was the Boeing 720 modified in a joint NASA/Federal Aviation Administration program to test a new jet

FAA

Fig. 5-11. *The world's largest UAV was this radio-controlled Boeing 720 which was crashed on the dry lake bed at Edwards AFB.*

fuel with much greater fire-retarding properties. In the Controlled Impact Demonstration (CID) program, the 720 was flown remotely until it was purposely crashed at Edwards AFB in December 1984 at a speed of 168 mph. Besides testing the fuel, the aircraft was instrumented to obtain data on crash forces as well as new seat designs and fireproofing materials.

6

Battlefield Robots

MILITARY personnel are interested not only in robotic unmanned air vehicles, but also robots that can perform missions on the ground. Like their counterparts in the sky, these robotic warriors can extend the capabilities of the individual soldier and are ideal for high-risk jobs. Many of the ground-based robotic vehicles are being developed to launch weapons, perform reconnaissance behind enemy lines, and operate in hazardous environments. Because much of the technology used in these robotic vehicles is similar to that found in unmanned air vehicles, it is of interest to look at some of these ground vehicles.

Use of High Technology

One of the major current robotic vehicle programs is part of the *Defense Advanced Research Projects Agency's* (DARPA) Strategic Computing Program. The overall object of the DARPA program is to develop machine or *Artificial Intelligence* (AI) technology. A key piece of hardware in this program is the *Autonomous Land Vehicle* (ALV) built to test ideas and technology under real-world conditions.

The eight-wheeled ALV developed by Martin Marietta in Denver, Colorado, is powered by a 6.2-liter diesel truck engine. Its 10-foot-long fiberglass body is filled with advanced computers and equally advanced sensors such as laser scanners and TV cameras. The 15,2000-pound ALV, which is completely self-contained, can operate independently without a human driver. This is done using a combination of artificial intelligence, sensors, and parallel processing computers that can handle multiple problems simultaneously. The TV cameras and laser scanners provide a picture of the terrain in front of the ALV. The computers, using their AI algorithms, process this information to command the ALV to maneuver autonomously. Since the vehicle can move only as fast as its computers can process the data, rapid parallel processing is a must. The ALV can also be remotely controlled from a command-and-control van.

The ALV is being used to develop hardware and software that will allow it to avoid obstacles and negotiate sharp turns, travel over rugged terrain, and use internally-stored terrain data to make long-distance trips. AI developments will be used to handle encounters with unforeseen terrain features. The ALV is not a prototype of any future robotic combat vehicle, but a test bed for developing and demonstrating AI, sensor, and computer technology that can be used in a multitude of future robotic vehicles that could eventually become operational with military forces.

Some of the ALV technology is already being used in the DARPA-sponsored *Advanced Ground Vehicle Technology* (AGVT) program. For example, the *Robotic Research Vehicle 3* (RRV-3) has been developed by General Dynamics Land Systems Division, Cadillac Gage Company, and Hughes Aircraft. A look at the AGVT shows how all these advanced technologies can be used in a vehicle that can either be controlled by a human operator with a minimum of training or travel autonomously. The RRV-3 is a 15,000-pound Cadillac Gage Commando Scout Vehicle that has been converted for both teleoperation (remote control) and autonomous operation.

Cadillac Gage Textron

Fig. 6-1. *This Cadillac Gage Commando Scout Reconnaissance Vehicle (RRV-3) can be teleoperated by remote control or run autonomously.*

In the teleoperated mode, an operator controls the RRV-3 from a mobile command and control center called the *Experimental Vehicle Analysis Laboratory* (EVAL) using a microwave communications link. The "driver" controls the RRV-3 with a steering wheel and foot pedal for the brake and throttle. These are rigged up so he actually gets a realistic "feel of the road" through feedback of steering torque and brake pedal pressure. Engine, suspension, and transmission sounds are also sent back to the driver. These features are a great help for a person with minimum training in operating the RRV-3. The driver views the scene

> The Driver's Control System (DCS), located in the Experimental Vehicle Analysis Laboratory (EVAL), provides the interface between man and machine. It is essentially a clone of the automotive controls found in the vehicle. The DCS processes operator commands (e.g., steering, braking, throttle) and then passes them on to EVAL's central command and control computer by way of the MIL-STD-1553B multiplexed data bus.

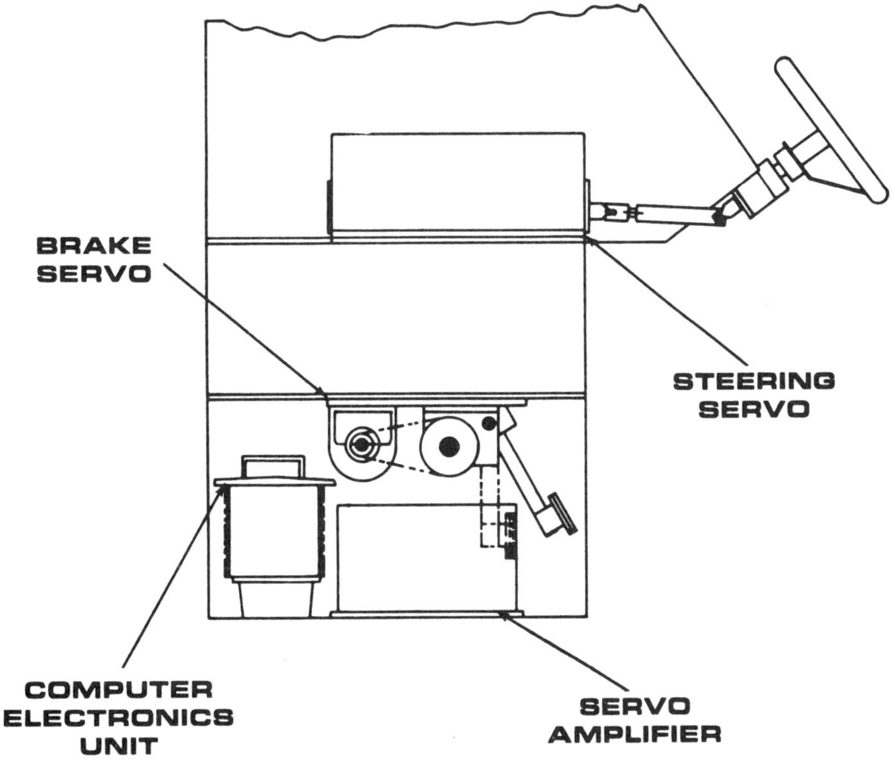

Fig. 6-2. Schematic of the Driver's Control System for the Cadillac Gage RRV-3.
Cadillac Gage Textron

outside on several color TV monitors that present information from video cameras and a thermal imaging sight aboard the RRV-3. There is a rearward looking camera that automatically displays the view to the rear when the RRV-3 is shifted into reverse. The thermal imaging equipment is used for night driving. Besides the TV monitors, there is a Multifunction Control and Display Panel for monitoring vehicle status and control functions, and a Symbolic Map system that provides the driver with real-time information on the RRV-3's location and heading.

The vehicle's commander is also provided with a redundant set of controls and monitors so he can not only do mission planning, but also act as a backup driver. However, the commander controls the vehicle by means of a joystick rather than a steering wheel, a feature included to obtain human factors information on the differences between the two methods of control. The commander also uses a reconnaissance mission module that features a high-resolution black-and-white TV camera with zoom capability for daytime operation and a thermal imager for nighttime use. The module used for reconnaissance along the route can be swung through a full 360 degrees.

But autonomous operation is the main thrust of the program. Since the autonomous capability is derived from the ALV program, an explanation of the AGVT system shows how the ALV handles autonomous navigation and control. The AGVT has equipment and algorithms that convert video images of the road into a digitized road "map." Among the important features of this map are the road edges that are used by the autonomous guidance system. The command and control computer aboard the RRV-3 converts the road edges into a reference trajectory using its software. The software then uses this reference trajectory for steering, braking, and throttle commands, which are sent to the RRV-3's servo control system. While the autonomous guidance and control equipment is located onboard the RRV-3, the vision processing equipment is located in the EVAL and information is transferred by microwave radio. (Remember, this is still a test bed.)

FMC is also using robotic technology in a tracked vehicle, a modified 113A2 armored personnel carrier. (By comparison, the Cadillac Gage vehicle has four wheels fitted with tires.) Besides the FMC's robotic vehicle, there is a remote command-and-control vehicle since this robot too has both teleoperated and autonomous capability. Computers are also used to convert sensor data from color TV and infrared cameras to control commands.

Robotic Soldiers

The soldier's job in combat is dangerous. Battlefield robots are one way to reduce the danger. One "robot warrior" concept is the *Teleoperated Mobile Antiarmor Platform*, or TMAP, being developed by the Grumman Corporation for the U.S. Army. The four-wheel drive all-terrain TMAP is about the size of an electric golf cart—roughly six feet long, four feet wide, and four-and-a-half feet tall. Constructed of lightweight and battle-resistant composite materials, the TMAP weighs only about 600 pounds. While small, it could still carry some potent weapons—for example, several tank-destroying TOW or Hellfire missiles, anti-

Grumman Corporation

Fig. 6-3. Artist's concept of the Grumman Teleoperated Mobile Antiarmor Platform (TMAP).

aircraft missiles such as the Stinger, plus a couple of machine guns or grenade launchers for self-defense.

The Grumman design uses four wheels laid out in a diamond pattern, with single wheels fore and aft and two side wheels mounted on a central axle. This layout provides better stability and maneuverability in rough terrain. For the same reasons, the TMAP has a segmented, articulated body. The operator could remotely control the TMAP from as far away as 2.5 miles; thus he could remain safely hidden from enemy fire. TV cameras and other sensors aboard the TMAP are used for control as well as locating targets and firing weapons. The operator would communicate with and control the TMAP either by radio or fiber-optic cables using a portable control unit in a backpack he carries. Martin Marietta Aero and Naval Systems has also proposed a similar TMAP vehicle.

The Fire Ant developed by Sandia National Laboratories is another robotic antiarmor weapon. The Fire Ant is a "one-shot" weapon that carries a single munition and is destroyed along with the enemy tank. Even though each Fire Ant

Fig. 6-4. The Martin Marietta version of the TMAP.

weapon might cost several thousand dollars to build, it would be able to destroy a multimillion-dollar tank—a pretty good economic tradeoff.

The Fire Ant can detect and destroy armored vehicles on the move at distances of up to 550 yards. The soldier operates the Fire Ant by radio control, although fiber-optic cables might also be used. The soldier only positions and aims the vehicle. The munition is fired automatically once the target enters the Fire Ant's aim point. The operator could override the automatic system. TV pictures from a small video camera mounted on the Fire Ant are displayed on a small screen on the remote control unit and the operator uses a joystick to accelerate, brake, and steer the vehicle; switches control the other functions. Once the vehicle has been positioned and aimed, the computer takes over to arm and fire the weapon using video information to determine target motion and predict the precise aim point.

There are two parts to the system: the robotic vehicle that carries the weapon and a small black-and-white video camera, and the remote controller and battery pack carried by one or two soldiers. For the prototype, a regular *All-Terrain Vehicle* (ATV) was modified and powered by a 125cc gasoline engine. Electrically driven actuators steer and brake the vehicle as well as operate the throttle and tilt the weapon platform. The weapon is a 22-pound copper slug that is propelled at

Sandia National Laboratories

Fig. 6-5. *The Fire Ant remotely controlled antitank weapon would destroy an enemy tank using the munition that looks like a searchlight seen in this photo.*

6600 feet per minute to distances of up to a third of a mile away. There is no gun barrel per se; the slug is fired from a device that looks like a spotlight on top of the vehicle.

To keep the costs of the systems expendable parts to a minimum, high-cost items such as infrared imagers, inertial guidance systems, and computers would be contained in the reusable control unit. One tactic is to use a larger (but still unmanned) teleoperated vehicle that could carry, tow, or lead four to ten smaller Fire Ants into battle. Alternately, many Fire Ants could be controlled from a single fixed command post. The Fire Ant vehicle could be used for reconnaissance, spotting and designating targets, and gathering intelligence behind enemy lines.

Another idea for a "kamikaze-type" weapon is the remotely controlled Sprinkler proposed by Universal Military Robot Corp. The $2000 Sprinkler is about four feet long and carries a machine gun. It could move among the enemy spraying bullets at the rate of 720 rounds per minute before it was either destroyed or ran out of ammunition.

Not all the combat robots would travel on wheels or tracks. Since about half the Earth's surface is not accessible to wheeled or tracked vehicles, some will have to do what "real" soldiers do—walk. Some time back, researchers at Ohio State and Carnegie-Mellon Universities and Odetics, Inc. developed a robot called ODEX that walked on six legs (called "articulators") that surrounded a cylindrical body. On ODEX's top is a video camera and an articulated arm with two "fingers" for manual tasks. ODEX can move at speeds up to eight mph and lift weights up to just over a ton—about five times the weight of the robot itself. ODEX can climb stairs or move over 33-inch-tall obstacles. The ODEX's brain comprises seven microprocessors.

Robots as Noncombatants

The Army is not the only military service interested in ground-based robots. For example, in time of war the Air Force must be able to quickly repair bomb-cratered runways even in a chemical or biological warfare environment. For this purpose, John Deere is developing its Rapid Runway-Repair Excavator that can fill craters and remove bomb debris. This device could be either manned or operated remotely from a safe distance.

MARVIN is a robot that was developed by the Air Force to work in hazardous environments. Not only does the four-foot MARVIN work like a human, it even looks like one with its molded plastic skin covering its sensitive electronics and computer. MARVIN could turn its head, move its arms, and grasp objects to lift them. However, instead of walking, MARVIN did roll on wheels.

The Air Force is also interested in robots to repair, refuel, and rearm aircraft in hazardous environments, as well as fight fires and repair damaged runways. Because of projected manpower shortages, a single human could operate perhaps a half dozen robots. In order to reduce the workload, the robots would be preprogrammed so they could autonomously perform routine tasks and would need operator assistance only when they ran into trouble.

The U.S. Navy is developing robots such as ROBART II to detect smoke, fire, flooding, hazardous gases, or even intruders aboard ships or on shore. The 50-inch-tall ROBART is made of plastic and fiberglass and uses a combination of microwave, ultrasonic, infrared, optical, and auditory sensors to do its job. One unique feature of ROBART is its voice synthesizer for communicating with

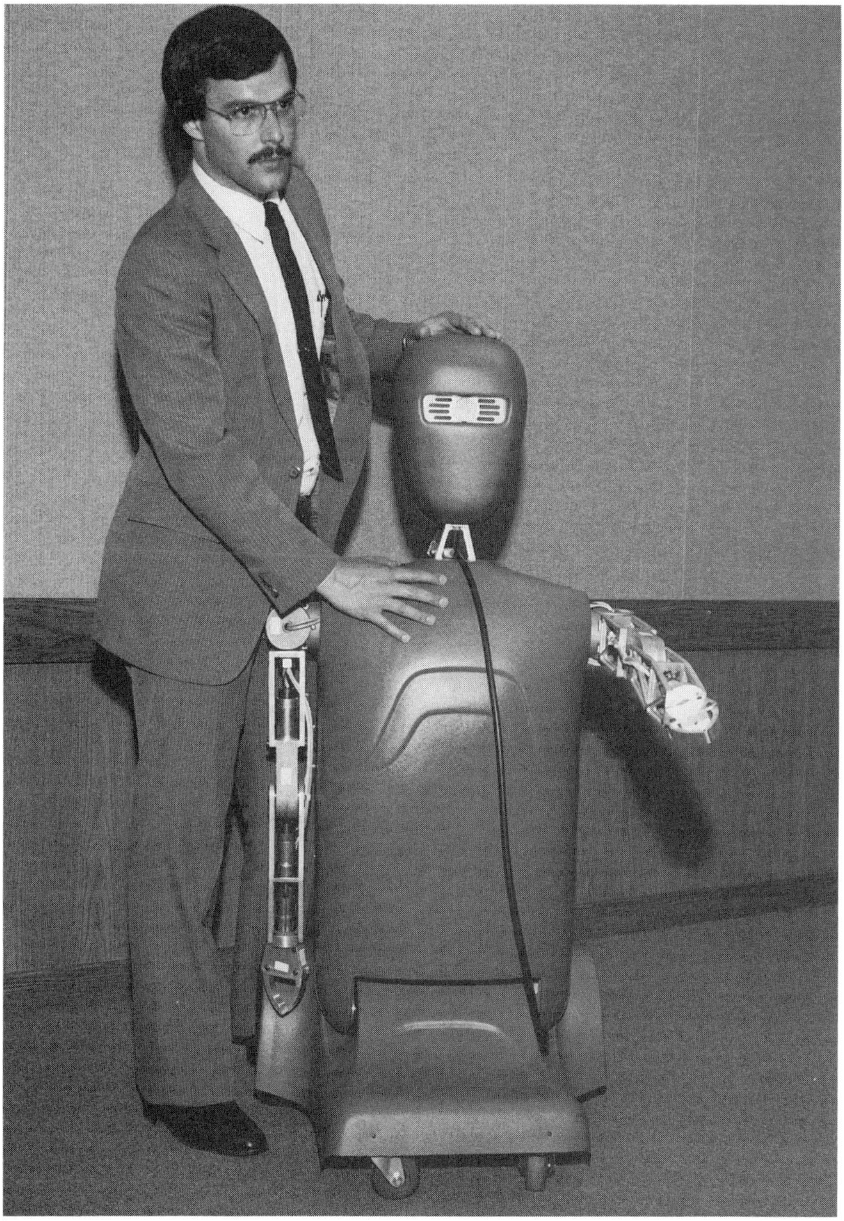

USAF

Fig. 6-6. Robots such as this USAF-developed Marvin could work in hazardous environments.

humans. When ROBART detects an intruder (its infrared sensor can detect human body heat), it could challenge the intruder verbally or summon human help.

7

UAVs in Civilian Clothes

UNMANNED air vehicles, as well as robotic vehicles on the ground, could be used for a variety of civilian jobs. However, for a number of reasons, few civilian-oriented UAVs have been developed and actually put into routine service.

Potential Civilian Jobs

Several years ago, NASA did an exhaustive study of civilian uses for remotely piloted vehicles. They identified 35 specific applications. Let's take a look at several of these applications, most of which are as attractive now as they were in the late 1970s.

As already proven on military missions, RPVs are a natural for surveillance of both small and large areas. Small-area surveillance could include looking for crimes, fires, or other emergencies in progress in a limited area such as a railroad yard, warehouse district, or industrial facility. Such an RPV would carry a low-light-level TV camera for work at night. The RPV would be controlled by a security guard, who would also monitor the TV picture. The RPV's flight could be preprogrammed with the operator having the ability to take over when he wanted a closer look. The RPV could also loiter over the scene of an emergency. A spotlight aboard the RPV and operated from the ground could illuminate the scene and a loudspeaker could allow the guard to speak to an intruder.

Large-scale surveillance RPVs could check for compliance with land-usage regulations and track pollution of streams and rivers. They could also be used to spot illegal dumping of waste and hazardous materials, thermal pollution from

Fig. 7-1. This robot would be used to inspect for leaks and damage in ventilating ducts and pipelines using the front-mounted light and video camera.

powerplants, and oil spills. In the case of the latter, RPVs could be very valuable in cleanup operations to help direct the placement of boats, skimmers, and containment booms. Oil slicks are much easier to see from the air. RPVs could be used to keep track of trucks and railcars carrying nuclear waste and other hazardous materials.

RPVs could also be used to cover vast forest or brushland areas to detect fires before they spread. By mapping forest and wildfires, observers could direct firefighting operations, locate hot spots, and keep track of changes in the spread of a fire so firefighting crews could be dispatched to the right location.

RPVs could fly over ice fields to assist icebreakers in their travels. In the Great Lakes, they could be used to locate ice-free channels and determine ice thickness to aid winter shipping. Photography done from an RPV could be used for precise land surveys for a variety of purposes. Large-area surveillance would be useful in search and rescue operations to locate survivors of aircraft crashes or disasters at sea.

Law enforcement agencies could use RPVs to monitor highway traffic conditions, assist ground officers in detecting and preventing criminal activities in progress, direct ground units to apprehend fleeing suspects, or perform search-and-rescue missions. Patrols of highways could locate accidents, motorists in trouble, stolen vehicles, or unsafe road conditions such as landslides and washouts. Borders could be patrolled for illegal aliens and drug smuggling. By monitoring the TV transmissions from several RPVs, a single officer could "cover" much more territory than he could from a patrol car. RPVs would be especially attractive in desolate areas of the country, where law enforcement agencies must provide surveillance of vast amounts of real estate with a minimum of personnel.

RPVs could be used for land management and locating resources such as oil and coal. They could also be used to locate schools of fish in the ocean and direct

Sandia National Laboratories

Fig. 7-2. *This* Sandia Interior Robot *(SIR) could be used for roving patrols inside a high-security building using a miniature television camera.*

fishing boats to make their harvest, as well as to detect illegal offshore fishing by foreign ships.

The study also identified what it called "linear patrol," which includes patrolling gas and oil pipelines to detect leaks and other hazardous conditions. By flying along a power line, an RPV operator could locate broken insulators, structural problems with electric towers, and hot spots such as those caused by overheated transformers.

RPVs could be used for spraying both crops and forests, and fire retardants could be dropped from RPVs during forest and brush fires. For crop spraying, the RPV would be remotely controlled while flying very close to the ground, eliminating danger to human pilots.

Another important role might be sampling the atmosphere. This includes both basic research and real-time detection of thunderstorms, hurricanes, and tornados. Pollution sampling in urban areas could provide information needed for smog alerts. In meteorology, the RPV could replace the weather balloon.

RPVs could be used to monitor ground sensors for detecting illegal border crossings or takeoffs and landings from remote airstrips suspected of being used by drug smugglers. More mundane sensors could be used to monitor corrosion of pipelines. Radio signals from emergency beacons aboard downed aircraft could be detected from RPVs.

Why So Few Civilian UAVs?

While UAVs are attractive for a number of civil roles, very few have been put into use even though most of the technology is already here. Indeed, many of the appropriate jobs are really civilian versions of military missions. For example, police surveillance for criminal activity is quite similar to spying behind enemy lines. Even the threat of "enemy" small arms fire is present in both cases.

Civilian UAVs would have to be rugged and reliable when used in bad weather, dust, heat, cold, and rough handling. They would have to be safe to fly and easy to operate by relatively inexperienced "pilots." Low initial and operating costs are also important. These are the very same characteristics that the military is looking for in *its* UAVs.

The main reason that civilian UAVs have not yet been developed to their full potential is, of course, the lack of funds. Most of the applications are in the public rather than the private sector. Civilian agencies with modest resources are not able to develop UAVs with the fervor of the military; nor do they (in most cases) even have the money to purchase "off the-shelf" UAVs and train their people to use them. This situation will probably not improve in the future, because federal, state, and local governments are strapped for funds and high technology items such as UAVs wind up on the bottom of their budget priorities. One bright spot is in the war on drugs. Here, not only can UAVs be cost-effective, but there may be sufficient funding available.

The private sector does not pick up on new aerospace developments that are still a bit risky; it usually waits for the government to do the developing and for the costs of such development to be absorbed. This probably means that civil RPVs are still many years away.

Environmental and safety concerns are two more potential major impediments to wide-scale civil use of UAVs. Some UAVs—especially the very low-cost ones—can be as noisy as radio-controlled model aircraft. (Some of them are, in fact, just modified R/C aircraft.) Many of their missions would require the craft to fly at low altitudes over heavily populated areas. The safety concerns are obvious. Other UAVs would fly in an already-crowded sky filled with commercial and general aviation aircraft. In both cases, the lack of a human pilot aboard could further compound the problem.

Liability and insurance would be other major concerns. Can civilian UAVs even operate in today's litigious society? Reportedly, about 70 percent of the cost of a new light airplane goes for product liability insurance. Because UAVs do not have a proven safety record, the problem could be even worse.

A final obstacle might be public acceptance. Would civilians—particularly American citizens—be willing to put up with robotic "spies in the sky," or would they view civil RPVs as an Orwellian nightmare?

Civilian UAVs Already Flying

Even with the problems listed above, there are civilian UAVs in operation—or, at least, they have been used in the past.

UAVs have been shown to be a very attractive alternative to space satellites. They are much less expensive and much more flexible. Here, UAVs must compete with *higher*-cost systems, rather than lower-cost alternatives (at least in the short term) such as a police officer in a patrol car. Furthermore, these UAVs fly at very high altitudes, away from many of the problems of safety and concerns for the environment.

NASA is particularly interested in UAVs as an alternative to high-cost satellites, as well as a replacement for sounding rockets for atmospheric research. In the 1970s, NASA developed its series of "Mini-Sniffer" RPVs. These were small, low-cost UAVs powered by a small piston engine; they were designed for such tasks as measuring air turbulence and atmospheric trace gases at altitudes up to 90,000 feet.

In the early 1980s, NASA converted a Volmer Sunfun powered hang glider for remotely controlled flight. The LRV-2 (*Low Reynolds Number Vehicle 2*) was designed for low-duration, high-altitude, and slow-speed flight. NASA demonstrated its ability to perform Earth resource research and relay communication signals, a job normally performed by much more expensive space satellites.

A more ambitious program is the NASA/Lockheed Solar *High-Altitude Powered Platform,* or Solar HAPP. The Solar HAPP would be a very large but lightweight unmanned aircraft that could stay aloft for up to a year. It would loiter at an altitude of about 65,000 feet, where winds are the calmest. While less power would be needed to overcome winds, the Solar HAPP would require extremely large wings to provide sufficient lift in the rarified atmosphere.

To keep weight down to two tons for an aircraft with a wingspan of over 300 feet would require construction of ultra-lightweight materials, such as a graphite epoxy structure covered with Mylar and Teflon. Strength is also important, because even at an altitude of 12 miles, there are still wind gusts and turbulence.

Lockheed

Fig. 7-3. The Solar HAPP (High-Altitude Powered Platform) could be used for a variety of very long duration civilian missions at very high altitudes, replacing much more expensive space satellites.

The Boeing Co.

Fig. 7-4. The Boeing Condor has set several altitude records for piston-powered aircraft.

Mollier International

Fig. 7-5. The flying-saucer-like Mollier Aerobot uses ducted fan propellers so it can take off and land vertically. This 30-inch-diameter version is powered by four 2-horsepower electric motors and could be used for surveillance tasks using a video camera.

The HAPP would still take off and land from the ground. The plan is to launch the craft on a calm day, taking up to four hours to spiral up to the desired altitude. During daylight, the HAPP's solar cells would capture sunlight and convert it to electric power. This power would be used directly during the day and also charge fuel cells for operation at night. The HAPP's wingtips would be folded down at night to improve its aerodynamics and thus reduce power requirements. Electric motors would power the slow-turning pusher propellers, which could push the HAPP at speeds of up to 90 mph.

NASA is interested in craft such as the HAPP for such jobs as agricultural monitoring, but the military applications in surveillance and communications are quite obvious.

Another UAV with both military and civilian application is the Boeing Condor. Officially, the Condor is the *High-Altitude, Long-Endurance* (HALE) autonomous unmanned aircraft. Flying at ultra-high altitudes for days at a time, the HALE could be used for such functions as border patrolling, mapping, reconnaissance/surveillance, and communications relay.

The Condor's structure is made of bonded composites and it has a 200-foot span (wider than a Boeing 747!) high-aspect ratio laminar-flow wing so it can fly very efficiently at high altitudes. It is powered by a liquid-cooled turbocharged six-cylinder engine. The Condor can fly totally autonomously from takeoff through landing according to a preprogrammed mission loaded into its computers prior to takeoff. The flight plan can be modified during flight via radio commands from the ground.

In 1989, the Condor broke the altitude record for piston-powered aircraft held for just over 50 years by an Italian Caproni biplane. The previous record (17,083 meters or 56,046 feet) had been set by Mario Pezzi. The Condor surpassed this record by climbing to 20,421 meters (66,980 feet) above sea level. The Condor also established a national record for altitude in horizontal flight for piston power, beating the 9,693-meter (31,792 feet) record set by American Stephen Oster in 1983.

Mollier International of Davis, California, builds some unique UAVs. Its Aerobots look like flying saucers with their ducted propellers and circular shape. The Aerobots come in a variety of sizes, from a 25-pound, 15-inch diameter model that can carry a 15-pound payload to an 1150-pound, 9-foot diameter version that can carry up to 750 pounds. Power comes from either an electric motor or a rotary gasoline engine that produces as much as 160 horsepower. One advantage of the Aerobots is their ability to take off and land vertically, as well as hover in place. The Aerobots are capable of performing such civilian jobs as security surveillance, fire spotting and suppression, traffic surveillance, pipeline/transmission/rail inspection, video news gathering, and inspection/surveying/exploring in remote or hazardous areas.

Civilian-oriented UAVs are not unique to the United States. Over in France the *Etablissement d'Etudes et de Recherches Meteorologiques* has used a series of rather simple RPVs since the mid-1970s to sample the atmosphere at altitudes up to about 16,000 feet. These EERM SAM-B/C/D RPVs are made of fiberglass and polystyrene and use a 1.6-horsepower 10cc single-cylinder engine. They are radio controlled in flight and are used to take measurements such as temperature, pressure, and humidity.

In Canada, Control Technologies, Inc., of British Columbia designed its Isis I and II RPVs for law enforcement, wildlife surveys, city planning, land development, search and rescue, marine studies, and forestry management. The Zenair Cricket, also built in Canada, was designed for pipeline patrolling using TV and still cameras; these were reportedly put into service in Qatar.

The British-built Eyrie UMAC/Merlin is an extremely small RPV, weighing 15 pounds at launch and, depending on the model, having a wingspan ranging from 7.5 to just under 11 feet. For silent operation, the craft is powered by one or two electric motors. Planning missions include aerial photography and television filming using either a 16mm or video camera. The control equipment is "off-the-shelf" R/C gear.

Index

V

W

X

Z

Other Bestsellers of Related Interest

THE PILOT'S AIR TRAFFIC CONTROL HANDBOOK—Paul E. Illman

The human elements of pilot and controller are the main focus of this new book by well-known aviation author Paul E. Illman. An understanding of what one expects of the other, and the responsibilities that belong to each are examined, answering important questions often asked by VFR pilots. Direct quotes and suggestions from ATC personnel are highlighted. 224 pages, illustrations. Book No. 2435, $16.95 paperback only.

THE PILOT'S RADIO COMMUNICATIONS HANDBOOK—3rd Edition—Paul E. Illman and Jay Pouzar

" . . . should have a spot on your bookshelf . . . "
—Private Pilot

" . . . time spent on this book is sure to make your flight smoother."
—Kitplanes

An updated edition of a popular handbook. Contains information on FAA rule changes regarding Mode C transponders, single-class TCA operations, and student pilot TCA training requirements. Current issues relating to the entire spectrum of VFR radio communications are addressed. Now, you can use even the busiest airports with confidence and skill. 240 pages, 61 illustrations. Book No. 2445, $15.95 paperback only.

MASTERING INSTRUMENT FLYING
—Henry Sollman with Sherwood Harris

Mastering Instrument Flying introduces an entirely new course designed from beginning to end to meet or exceed the Instrument Flight Test standards recently published by the FAA. The elements, techniques, procedures, and tolerances of instrument flight are addressed in precise detail. Illustrated information on how to prepare for the instrument flight test is provided, and additional advanced procedures not specifically required for the instrument rating are covered. 336 pages, 256 illustrations. Book No. 2433, $18.95 paperback only.

AVOIDING COMMON PILOT ERRORS: An Air Traffic Controller's View—John Stewart

A pilot, flight instructor, and air traffic controller of 20 years' experience cites examples of recurring pilot error. Improper training, lack of preflight preparation, poor communication skills, and confusing regulations are among the problems discussed. 240 pages, illustrated. Book No. 2434, $16.95 paperback only.

IMPROVE YOUR FLYING SKILLS: Tips from a Pro—Donald J. Clausing

Learn firsthand the professional attitudes, flying standards, and everyday procedures practiced by airline and corporate aircraft captains. Leading off with an overview of the basics of flying VFR and IFR, the author gives in-depth coverage of all the things that make up advanced, no-nonsense airmanship. Among the topics covered: flight planning, cruise control, types of approaches, weather flying, filing IFR, radio procedures, and "BILAHS" (Briefing, IFR, Log, Alternate, Hazardous weather). 224 pages, 42 illustrations. Book No. 3328, $14.95 paperback, $24.95 hardcover.

THE ILLUSTRATED GUIDE TO AERODYNAMICS—Hubert "Skip" Smith

If you've always considered aerodynamic science a highly technical area best left to professional engineers and aircraft designers . . . this outstanding new sourcebook will change your mind! Smith introduces the principles of aerodynamics to everyone who wants to know how and why aircraft fly . . . but who doesn't want to delve into exotic theories or complicated mathematical relationships. 240 pages, 232 illustrations. Book No. 2390, $17.95 paperback only.

GOOD TAKEOFFS AND GOOD LANDINGS
—Joe Christy

This reference thoroughly examines single-engine aircraft takeoffs and landings, and the critical transitions accompanying each. The author stresses that every pilot must continually evaluate ever-changing factors of wind, air pressure, precipitation, traffic, temperature, visibility, runway length, and braking conditions. *Good Takeoffs and Good Landings* belongs on every pilot's required reading list. 192 pages, 70 illustrations. Book No. 2487, $14.95 paperback only.

THE ART OF INSTRUMENT FLYING
—J. R. Williams

This book addresses all elements of IFR flight using situations derived from the author's actual flight experiences. Each subject inherent to safe IFR operations is covered: instrument approaches and departures, weather, airspeed/altitude control, radio communications, charts, IFR flight planning, and spatial disorientation. 272 pages, 85 illustrations. Book No. 2418, $16.95 paperback only.